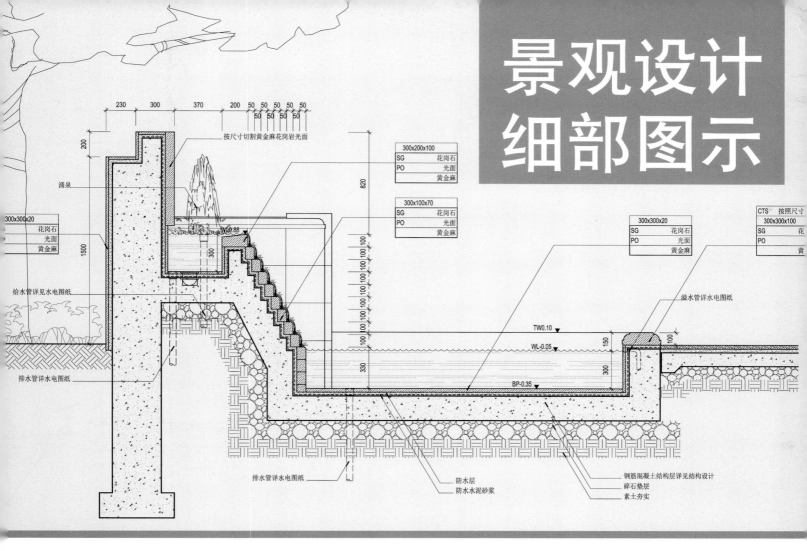

景观设计细部图示

《景观设计细部图示》为系列丛书，全套共六本，收集了泛亚国际成立30年来经典项目的CAD图纸，书中内容涉及景观工程的各个细部，包括水景、围墙、铺地结构、道牙、驳岸、扶手、花钵、排水台阶、泳池、隐形消防车道等22种常见景观类设计图纸，配以专业的规范、标准详解，融合专业性、学术性与知识性于一体。

The series book of Detail Diagram of Landscape Design including 6 volumes collects the CAD drawings of classical works of EADG in recent 30 years. Referring to every detail part of landscape design like water feature, fence, feature paving, kerb, pond-edge, railing, drainage, swimming pool and contact fire lane, etc, the series presents common drawings of 22 types of landscape design and construction. With professional text of design regulations and standards, the series will be an integration of professional work, academic achievements as well as practical knowledge.

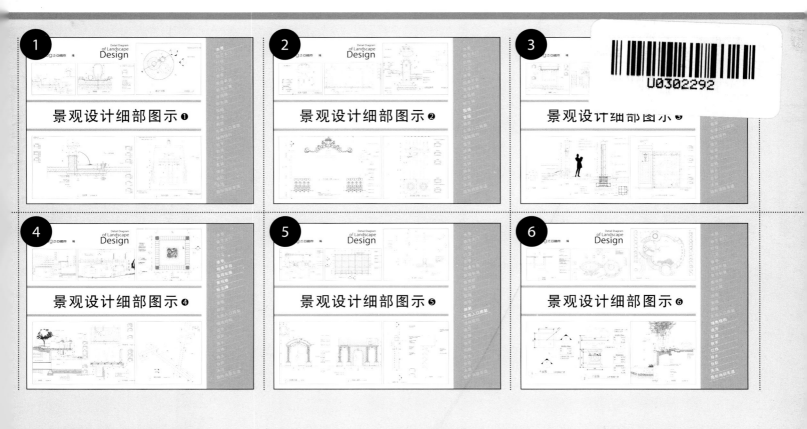

所有人名按字母顺序排列 Name in alphabetical order

Publisher出版人 Sun Xueliang
Distributor发行人 Sun Xueliang
Committee编委 Alexander Blakely, Altaf Master, Andre Kikoski, Bjarke Ingels, Brad Tomecek, Christopher Herr, Davide Macullo, Diego Burdi, Dwayne Oyler, Evzen Novak, Fernando Forte, James Mary O'Connor, Jenny Wu, Kevin Sietmann, Lourenço Gimenes, Mack Scogin, Masaru ITO, Matthias Altwicker, Merrill Elam, Paul Filek, Rand Elliott, Robert Oshatz, Rodrigo Marcondes Ferraz, Shigeo YOKOTA, Stefano Tordiglione, Steven Holl, Vlado Valkof

Editor-in-Chief执行主编 石莹 Erica
Marketing Manager市场总监 林佳艺 Rita
Editor-in-Charge编辑部主任 张晓华 Kitty
English Editors英文编辑 毛玲玲 Marian 石莹 Cindy
王培娟 Joan
Art Editor美术编辑 夏金梅 Candy
Contributing Editors 特约编辑 Aileen Forbes-Munnelly, Andre Kikoski, Brad Tomecek, Cateline Leung, Catherine Meng, Christopher Boyce, Daria Pahhota, Dwayne Oyler, James Mary O'Connor, Julia van den Hout, Marcellus Lilley, Margaret Fletcher, Mayumi MORINAGA, Michelle Jameson, Robert Oshatz, Rodrigo de Moura, Shigeo YOKOTA, Sonia Germain, Vlado Valkof

Price 定价 USD 29 RMB 98
Add:上海市虹口区天宝路886号天宝商务楼3楼
3rd Floor, NO. 886 Tianbao Road, Hongkou District, Shanghai, China
Post Code: 200086
Tel: 86-21-65018097-378
Fax: 86-21-65018097-188
E-mail: info@internationalnewarchitecture.com
Website: www.internationalnewarchitecture.com

Media Link媒体支持

X-COLOR. new landscape AIA

Supportive Companies & Associations 理事单位

EADG泛亚国际

图书在版编目（CIP）数据

国际新建筑. 12 / 凤凰空间•上海编. -- 南京：江苏人民出版社，2012.12
ISBN 978-7-214-08955-7

Ⅰ. ①国… Ⅱ. ①凤…Ⅲ. ①建筑设计－作品集－世界－现代 Ⅳ. ①TU206

中国版本图书馆CIP数据核字(2012)第279452号

国际新建筑12
策划编辑：石 莹 林佳艺
责任编辑：刘 焱
特约编辑：石 莹
出版发行：凤凰出版传媒股份有限公司
　　　　　江苏人民出版社
　　　　　天津凤凰空间文化传媒有限公司
邮　　编：210009
印　　刷：上海当纳利印刷有限公司
开　　本：965毫米×1270毫米 1/16
印　　张：12
字　　数：96千字
版　　次：2012年12月第1版
印　　次：2012年12月第1次印刷
书　　号：ISBN 978-7-214-08955-7
定　　价：98.00元
销售电话：022-87893668
网　　址：www.ifengspace.com
（本书若有印装质量问题，请向销售部调换）

《国际新建筑》
International New Architecture

卷首语

美国建筑师协会（AIA）成立于1857年，至今已有155年的历史，在世界建筑界已形成一股举足轻重的力量。这个跨越了3个世纪的协会伴随着建筑学的发展，见证了建筑领域内一幕幕华丽的转变与质的飞跃。毫不夸张地说，在这短短的155年当中，建筑理论和实践的发展远远超过了过去上千年艰难的摸索。作为一个有影响力的建筑奖项，AIA的历史也印证了人类建筑的轨迹。

编辑们此次为大家精选了部分2012年AIA获奖作品，从各个角度展现当代建筑的水准。耳闻不如一见，真正的大师给我们留下的不仅仅是响亮的名号，更是浮华背后充满思想与智慧的作品。书中每一页都流露着令人感动、心动，甚至是激动的灵魂。一见倾心，再见难忘——这就是大作的力量！

书面的轻描淡写远不足以呈现建筑设计的魅力，编辑们只愿由此为广大读者推开通向人类建筑未来的一扇窗，愿世界建筑步步向前，深入世界每一个角落，将它的美丽带进人的内心，带入未来。

American Institute of Architects (AIA), which was established in 1857, has seen a history of 155 years till now. It comes across three centuries and gradually occupies an important role in architecture, which develops dramatically fast in this period. Actually, the advancement of architectural theory and practice has surpassed those of the last difficult millennium. As an influential award presenter, AIA has been tracing about the career of each excellent architect, and even human architecture history….

Words are but wind. You'll be deeply impressed by the rich thought and wisdom of master architects when reading the articles for yourself. We cover several 2012 AIA Award projects in this issue, showing the achievement of architecture in recent years. You'll feel dissolved and affectionate, even be excited. Love you once, love you twice—that's the power of masterpieces—never just an undeserved reputation!

The articles are just a delicate and partial touch on architecture. We'd like to encourage talented architects and designers to contribute more to the world by this way. We wish to see the bright future of architecture, spreading culture everywhere, enriching people's soul.

2012 AIA Awards Recipients
——2012 AIA 获奖作品精选——

Contents 目录

专题 Feature

006 克努特哈姆森博物馆
Knut Hamsun Center

受到哈姆森作品的启发,Steven Holl建筑事务所设计了一个"空小提琴盒"造型的楼层,阳台的设计看起来仿佛"女孩儿挽起袖子正在擦拭黄色的窗户"。
Inspired by passages of Hamsun's texts, there is an "empty violin case" deck, while a viewing balcony is like the "girl with sleeves rolled up polishing yellow panes".

020 城市海洋科学博物馆
Cité de l'Océan et du Surf

建筑的造型由立体概念"苍穹下/深海处"演化而来。内凹的"苍穹下"构建了主要的外部广场,"海洋广场"面向蓝天和大海,遥望远方的天际线。凸起的天花板结构构筑起了"深海处"的展示空间。
The building form derives from the spatial concept "under the sky" / "under the sea". A concave "under the sky" shape forms the character of the main exterior plaza, the "Place de l'Océan", which is open to the sky and sea, with the horizon in the distance. A convex structural ceiling forms the "under the sea" exhibition spaces.

030 8字形大楼
8 House

8字形大楼的中心被一分为二,形成两个不同的空间,大楼的中心主要是500平方米的公用设施。
The bow-shaped building creates two distinct spaces, separated by the center of the bow which hosts the communal facilities of 500 m².

044 盖茨电脑科技中心和希尔曼未来发电技术中心
Gates Center for Computer Science and Hillman Center for Future Generation Technologies

对于项目最好的诠释是:一个能凸显卡内基梅隆和电脑科学学院在世界级学术殿堂中的显著地位的建筑。
An architecture that represents Carnegie Mellon University's and the School of Computer Science's exceptional status among the world's leading academic institutions is best informed from within the project's own situation.

052 古根海姆博物馆的莱特饭店
The Wright at the Guggenheim Museum

雕塑的形态营造出了外翻的天花板。浪状的墙壁变成了舒适的座椅,弧形酒吧和公共餐桌使空间更具动感。玩味的造型使游客获得多样的体验。
The sculptural forms create a flared ceiling. The undulating walls become comfortable seating. The arced bar and communal table animates the space. The playfulness of these forms offers a dynamic experience for visitors.

058 费耶特维尔2030:交通之城
Fayetteville 2030: Transit City Scenario

2030交通之城方案是一个富足项目,因为在这个方案下,每家每户的交通成本将降到年收入的16%。
The 2030 Transit City Scenario plan is a prosperity-building program since the average household in rail transit cities spent 16% of its annual income on transportation.

072 Grangegorman总体规划
Grangegorman Master Plan

该设计将为校园提供世界一流的创新型便利设施,并通过现代手法将传统的学院建筑风格进行完美诠释。
The design offers world-class, innovative facilities for both DIT and HSE, enhancing their identity and image by employing a contemporary interpretation of traditional collegiate quads.

080 金属盾包裹的房子
Shield House

该项目将一个高高的、弯曲细长的流通空间与矩形的居住空间并置。
This urban infill project juxtaposes a tall, slender curved circulation space against a rectangular living space.

建筑 Architecture

086 圣西拉斯小学
St Silas Primary School

096 成城庭园
SEIJO CORTY

104 经堂庭园
KYODO CORTY

116 中央舞台——南加州建筑学院毕业作品展
Center Stage: SCI-Arc Graduation Pavilion

124 锦州新区医疗中心
Jinzhou New Area Medical Center

128 Natura圣安德烈分店
Natura's Showroom Santo André

138 张氏别墅
Cheung Residence

142 Jansen总部大楼：树立愿景
Jansen Campus: Building A Vision

156 奥克兰交通科技博物馆飞机展示厅
MOTAT Aviation Display Hall

162 韦斯住宅
Weiss Residence

室内 Interior

168 独具特色的办公空间
ImageNet Houston

174 飞檐魅影
HIMEJI FORUS

178 暗黑潮流—Roar品牌旗舰店
Roar (Daikanyama)

182 apple & pie童鞋专卖店
apple & pie Children-shoe Boutique

188 布朗托马斯珠宝店
Brown Thomas Jewellery

阳台如小女孩挽起的衣袖
The balcony looks like the sleeves of a little girl

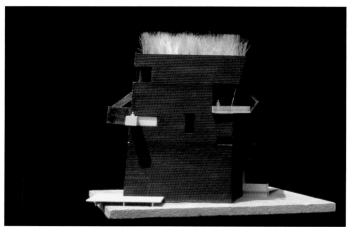

| Steven Holl Architects |

Knut Hamsun Center
克努特哈姆森博物馆

2012 Gold Medal Recipient   2012金奖

为了表彰史蒂文·霍尔对建筑的杰出贡献，他的名字将被刻在华盛顿AIA总部大厅的荣誉墙上。
霍尔之前获得的奖项包括：
• 2个英国皇家建筑师协会国际奖
• 9个美国建筑师协会国家荣誉奖

In recognition of Steven Holl legacy to architecture, his name will be chiseled into the granite Wall of Honor in the lobby of the AIA headquarters in Washington, D.C.
Holl's previous awards include:
• 2 RIBA International Awards
• 9 AIA National Honor Awards

外部设计思路手绘图
Concept sketch exterior

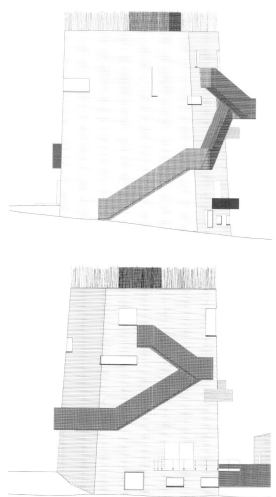

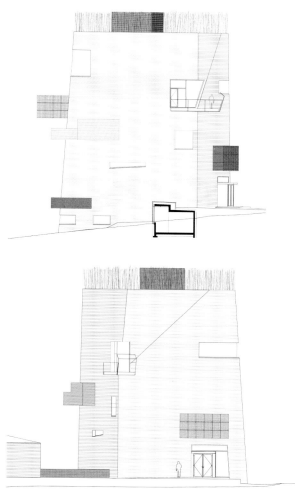

外部阳台视野极好，可以眺望山水相接的美景
The exterior balcony overlooking mountains and waters

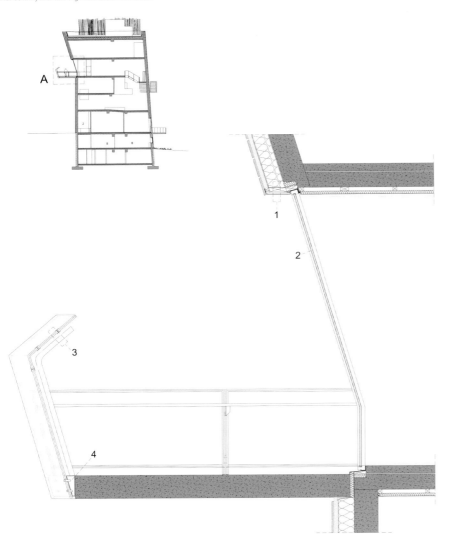

1. 1/8" (3mm) Thick sheet steel 5/8"(15mm) plywood panel, frame of 2 1/8x7 7/8" (200mm)mineral wool insulation
2. Aluminium window with black finish and 1/8-5/8-1/8"(4/15/4mm) double glazing
3. 1/8 (3mm) Thick galvanized steel T-profile with welded steel plate and threaded steel pins supporting 3/8 + 3/8" (10+10mm) trasparent yellow glass
4. 1/8 (3mm) thick box-shaped galvanized steel plate closure and fixing
5. 4x4" (100x100mm) steel box profile fastening projecting steel frame to reinforced concrete edging
6. Cladding in 1/8" (2.5mm) thick steel plates, frame supporting cladding in HEB 100 steel beams
7. 5/16-1/8-5/16" (8/7/8mm) double glazing assembly in frame at slab edge
8. 1/16" (2mm) thick shaped sheet finish
9. 4x9 1/8" (100x230mm) steel angle profile
10. floor in 4" (100mm) blck concrete with ground finish, , 1 1/4"(30mm) board insulation, projecting 7 7/8"(200mm) reinforced concrete slab, 4"(100mm) board insulation, 2x4" (50x100mm) steel C-Profiles fastening 1/8"(2.5mm) brass grille cladding
11. 1/8-5/8-1/8" (4/16/4mm) wood double glazing assembly
12. interior sill in 1/8" (3mm) sheet brass, 5/8"(16mm) plywood panel, 5/8x1 5/8"(15x40mm) wood batten, 1 5/8x2 5/8" (40x65 mm) steel C-Profile sandwiching 2 5/8"(65mm) board insulation, 1/8" (3mm) sheet steel installation space
13. Canopy roof in 1/8"(3mm) sheet brass, 3 1/8"(80mm) max H board insulation on 17 degree slope , 5/8" (16mm) gypsum board, 1 5/8x4 1/2 (40x115mm)steel C-Profile sandwiching 4 1/2"(115mm) board insulation, 1/8"(3mm) sheet steel, false ceiling in removable 1/8"(3mm) brass panels
14. 1/8" (3mm) Sheet brass, 5/8"(16mm) gypsum board, 1 3/8x2"(35x50mm) wood, 2 3/8"(60mm) board insulation, 2 3/4x7 1/8" (70x180mm) steel C-Beam, gutter, IPE 120 steel beam
15. Steel entrance door with 1/16-5/8-1/16" (5/15/5mm) double glazing
16. Exterior walkway in 2x108 1/4"(50x2750mm) wood, 9 3/8x3 1/8"(250x80mm) wood joists with pair of 3/8x4 3/4"(10x120mm) bolted steel plates securing beams to reinforced concrete foundation, gravel bedding
17. 1 1/2" (40mm) Galvanized steel grill, 4"(100mm) concrete, 2"(50mm) board insulation, 11 7/8"(300mm) reinforced concrete slab
18. 6 1/4x6 1/4"(160x160mm) steel angle profile
19. 1 5/8"(40mm) Thick door with 3 1/2x2 5/8"(90x65mm) steel frame with double glazing
20. Projecting walkway roof formed by 4x1"(100x25mm) untreated dedar wood, (30X30mm) wood beams, 1/8"(3mm) sheet steel, 5/16"(8mm) wood panel, 5 1/2"(140mm) max H wood beam, 1 3/8x2"(35x50mm)wood rafters, false ceiling in 1" (25mm) cedar wood panels
21. Gutter
22. External wall in 4x1"(100x25mm) untreated cedar wood, frame of 4"(100mm)wood, 4x1"(100x25mm) untreated cedar wood
23. 4x1"(100x25mm) untreated cedar wood frame of 1 1/4x1 1/4" (30x30mm)wood, 7 7/8" (200mm) reinforced concrete slab

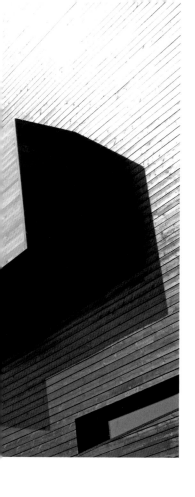
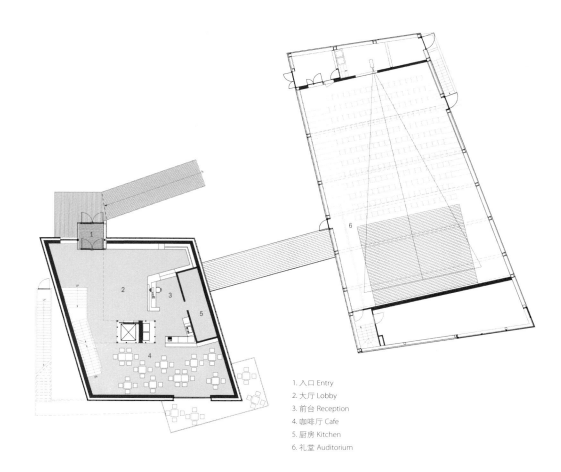

1. 入口 Entry
2. 大厅 Lobby
3. 前台 Reception
4. 咖啡厅 Cafe
5. 厨房 Kitchen
6. 礼堂 Auditorium

克努特·哈姆森是20世纪挪威最具创意的作家，同时是诺贝尔文学奖获得者，在他的第一本小说《饥饿》中创造了一种新的写作手法。这座纪念哈姆森的博物馆位于北极圈之上的哈马略，普瑞斯泰德村庄附近，与作者成长的农场毗邻。占地2 508平方米的博物馆囊括了展示区、图书馆及阅览室、咖啡馆和博物馆礼堂在内的几部分建筑。受到哈姆森对于人性复杂性探索的影响，被看做是一个建筑在空间和光线交错中的典型精神的浓缩，作为建筑体系中哈姆森的缩影。受到哈姆森作品的启发，Steven Holl建筑事务所设计了一个"空小提琴盒"造型的楼层，阳台的设计看起来仿佛"女孩儿挽起袖子正在擦拭黄色的窗户"。

该博物馆的设计理念"建筑等于人体：战场上的无形力量"由内而外的彰显出来。外部的木结构被隐藏着的冲击力量刺穿表面。建筑体的"脊梁"为中央电梯，由带孔的黄铜装饰而成。内部涂成白色的水泥板结构被日光的对角光线照亮，据计算，光线在每年特定日子会反弹并穿过该区域。柏油黑木外层影射挪威的中世纪木板教堂，顶楼花园上的竹制长槽模仿的是挪威传统草皮房顶。

礼堂与主楼通过一条底层的大厅走廊相连，这一设计充分利用了自然地形，让自然光线照进室内。

内部镂空墙壁
Interior perforating walls

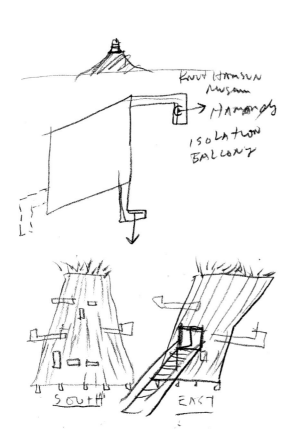

错落有致的内部空间
Interior space

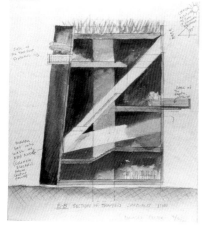
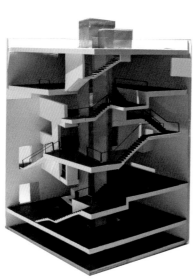
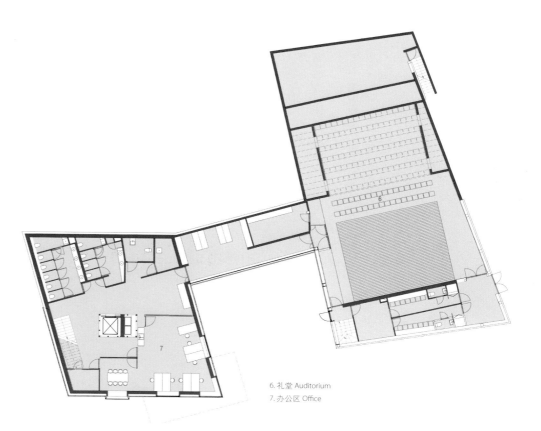

6. 礼堂 Auditorium
7. 办公区 Office

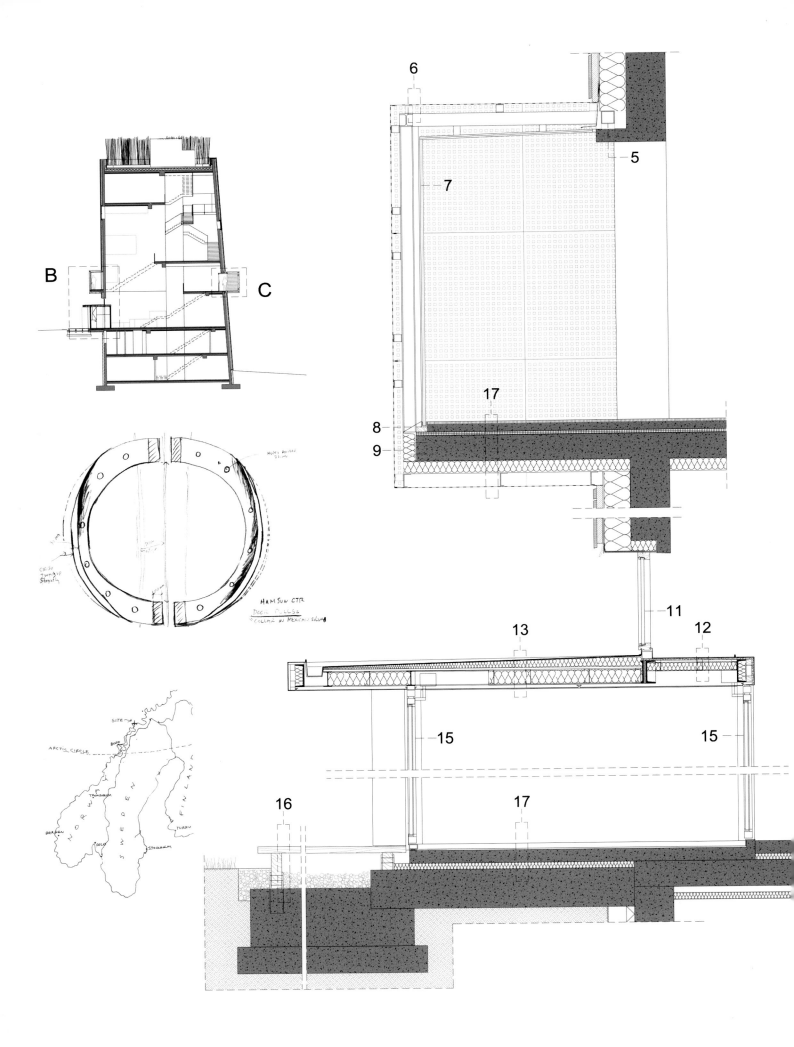

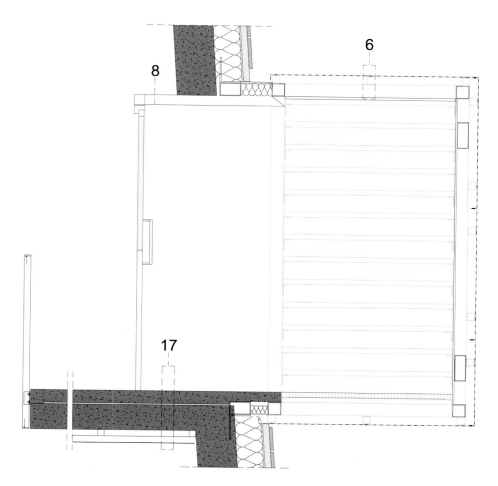

C

1. 1/8" (3mm) Thick sheet steel 5/8"(15mm) plywood panel, frame of 2 1/8x7 7/8" (200mm) mineral wool insulation
2. Aluminium window with black finish and 1/8-5/8-1/8"(4/15/4mm) double glazing
3. 1/8 (3mm) Thick galvanized steel T-profile with welded steel plate and threaded steel pins supporting 3/8 + 3/8" (10+10mm) trasparent yellow glass
4. 1/8 (3mm) thick box-shaped galvanized steel plate closure and fixing
5. 4x4" (100x100mm) steel box profile fastening projecting steel frame to reinforced concrete edging
6. Cladding in 1/8" (2.5mm) thick steel plates, frame supporting cladding in HEB 100 steel beams
7. 5/16-1/8-5/16" (8/7/8mm) double glazing assembly in frame at slab edge
8. 1/16" (2mm) thick shaped sheet finish
9. 4x9 1/8" (100x230mm) steel angle profile
10. floor in 4" (100mm) blck concrete with ground finish, , 1 1/4"(30mm) board insulation, projecting 7 7/8"(200mm) reinforced concrete slab, 4"(100mm) board insulation, 2x4" (50x100mm) steel C-Profiles fastening 1/8"(2.5mm) brass grille cladding
11. 1/8-5/8-1/8" (4/16/4mm) wood double glazing assembly
12. interior sill in 1/8" (3mm) sheet brass, 5/8"(16mm) plywood panel, 5/8x1 5/8"(15x40mm) wood batten, 1 5/8x2 5/8" (40x65 mm) steel C-Profile sandwiching 2 5/8"(65mm) board insulation, 1/8" (3mm) sheet steel installation space
13. Canopy roof in 1/8"(3mm) sheet brass, 3 1/8"(80mm) max H board insulation on 17 degree slope , 5/8" (16mm) gypsum board, 1 5/8x4 1/2 (40x115mm)steel C-Profile sandwiching 4 1/2"(115mm) board insulation, 1/8"(3mm) sheet steel, false ceiling in removable 1/8"(3mm) brass panels
14. 1/8" (3mm) Sheet brass, 5/8"(16mm) gypsum board, 1 3/8x2"(35x50mm) wood, 2 3/8"(60mm) board insulation, 2 3/4x7 1/8" (70x180mm) steel C-Beam, gutter, IPE 120 steel beam
15. Steel entrance door with 1/16-5/8-1/16" (5/15/5mm) double glazing
16. Exterior walkway in 2x108 1/4"(50x2750mm) wood, 9 3/8x3 1/8"(250x80mm) wood joists with pair of 3/8x4 3/4"(10x120mm) bolted steel plates securing beams to reinforced concrete foundation, gravel bedding
17. 1 1/2" (40mm) Galvanized steel grill, 4"(100mm) concrete, 2"(50mm) board insulation, 11 7/8" (300mm) reinforced concrete slab
18. 6 1/4x6 1/4"(160x160mm) steel angle profile
19. 1 5/8"(40mm) Thick door with 3 1/2x2 5/8"(90x65mm) steel frame with double glazing
20. Projecting walkway roof formed by 4x1"(100x25mm) untreated dedar wood, (30X30mm) wood beams, 1/8"(3mm) sheet steel, 5/16"(8mm) wood panel, 5 1/2"(140mm) max H wood beam, 1 3/8x2"(35x50mm)wood rafters, false ceiling in 1" (25mm) cedar wood panels
21. Gutter
22. External wall in 4x1"(100x25mm) untreated cedar wood, frame of 4"(100mm)wood, 4x1"(100x25mm) untreated cedar wood
23. 4x1"(100x25mm) untreated cedar wood frame of 1 1/4x1 1/4" (30x30mm)wood, 7 7/8" (200mm) reinforced concrete slab

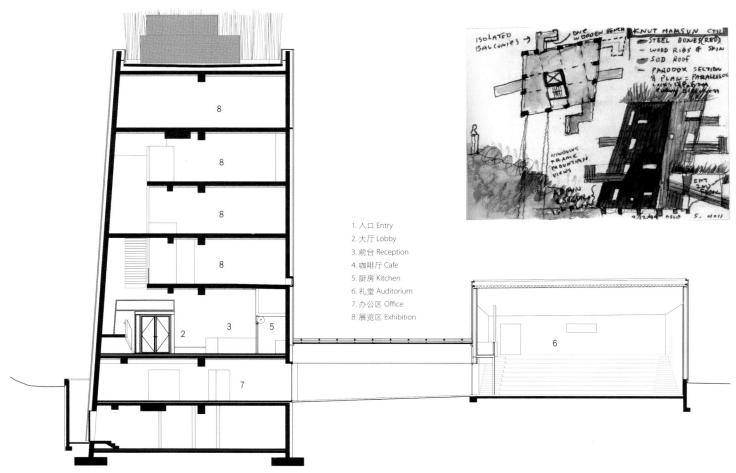

1. 入口 Entry
2. 大厅 Lobby
3. 前台 Reception
4. 咖啡厅 Cafe
5. 厨房 Kitchen
6. 礼堂 Auditorium
7. 办公区 Office
8. 展览区 Exhibition

Knut Hamsun, Norway's most inventive twentieth-century writer and recipient of the Nobel Prize in Literature, fabricated new forms of expression in his first novel Hunger. This center dedicated to Hamsun is located above the Arctic Circle by the village of Presteid of Hamarøy, near the farm where the writer grew up. The 2,508 m² center includes exhibition areas, a library and reading room, a café, and an auditorium for museum and community use. Influenced by Hamsun's explorations of the intricacies of the human mind, the building is conceived as an archetypal and intensified compression of spirit in space and light, and as a Hamsun character in architectonic terms. Inspired by passages of Hamsun's texts, there is an "empty violin case" deck, while a viewing balcony is like the "girl with sleeves rolled up polishing yellow panes".

The concept for the museum, "Building as a Body: Battleground of Invisible Forces," is realized from inside and out. The wood exterior is punctuated by hidden impulses piercing through the surface. The spine of the building body, constructed from perforated brass, is the central elevator. The board form concrete structure with stained white interiors is illuminated by diagonal rays of sunlight calculated to ricochet through the section on certain days of the year. The tarred black wood exterior skin alludes to Norwegian Medieval wooden stave churches and on the roof garden, long chutes of bamboo refer to traditional Norwegian sod roofs.

The auditorium is connected to the main building via a passageway accessed through the lower lobby, which takes advantage of the natural topography, allowing for natural light along the circulation route.

博物馆地处山水之间，风光无限
The museum is located between mountains and waters

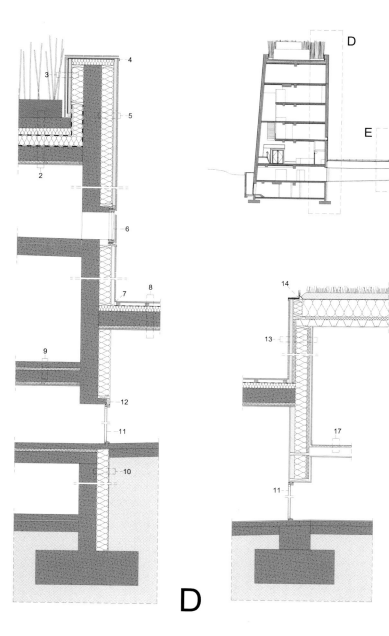

1. 11 7/8" (300mm) Concrete base for bamboo
2. Roof formed by 5 7/8"(150mm) concrete, waterproofing membrane, 5 1/8"(130mm) board insulation on 17 degree slope, waterproofing membrane, 7 7/8"(200mm) board insulation, vapour barrier, 9 3/8"(250mm) reinforced concrete slab, false ceiling with frame in 2x2"(50x50mm) wood with 2"(50mm) board insulation, 1'(25mm) wood and fibrocement acoustic panel
3. Edge cladding consisting of 4 1/2x1"(115x25mm) wood boards stained black, 2x2"(50x50mm) wood frame, waterproofing membrane, 5/8"(16mm) asphalt panel, 7 7/8"(200mm) board insulation
4. 1/8"(3mm) steel flashing
5. Cladding in 4 1/2x1"(115x25mm) wood boards stained black, 2"(50mm) thick wood beams, 3/8"(10mm) asphalt board, frame of 2x7 7/8" (50x200mm) wood sandwiching 7 7/8"(200mm) mineral wool insulation, 11 7/8" (250x300mm) reinforced concrete edge
6. 1/8-5/8-1/8"(4/15/4mm) wool double glazing assembly
7. 1/8" (3mm) Thick steel flashing
8. Connecting corridor roof in 1"(25mm) untreated oak panels, frame of 3 1/2x2"(90x50mm)wood, waterproofing membrane, 7 7/8"(200mm) board insulation, 9 3/8"(250 mm) reinforced concrete slab, 2"(50 mm) board insulation, 2"(50 mm) wood and fibrocement acoustic panel interior finish
9. Floor in 4"(100mm) black concrete with ground finish, 2" (50mm) board insulation, 7 7/8"(200mm) reinforced concrete fondation
10. 3/8"(10mm) drainage panel protecting foundation, 5 7/8"(150mm) board insulation, 11 7/8" (300mm) reinforced concrete foundation
11. Steel door
12. 3/8"(10mm) white painted gypsum board
13. Finish in 4 1/2x1"(115x25mm) pine boards stained black, 2"(50mm) thick wood beams, 3/8"(10mm) asphalt board, 1 3/4"(45mm) board insulation, 7 7/8"(200mm) mineral wool, insulation,waterproofing membrane, 1 3/4(45mm) board insulation, 5/8"(15mm) cabling spale, rendered 5/8"(15mm) birch wood board
14. 1/8"(3mm) steel flashing on 5 1/8x5 1/8"(130x130mm) steel angle profile
15. Green roof over auditorium formed by sedum, earth, triple waterproofing membrane, 11 7/8"(300mm) board insulation,waterproofing membrane, 2"(50mm) board insulation, 4 3/8"(110mm) corrugated sheet steel painted black
16. Gallery with 3/8"(10mm) birch wood panel finish, structure in HEB 100 steel beams
17. Gallery floor in 3/8"(10mm) cork on 3/4"(20mm) corrugated steel sheeting, IPE 160 steel beam

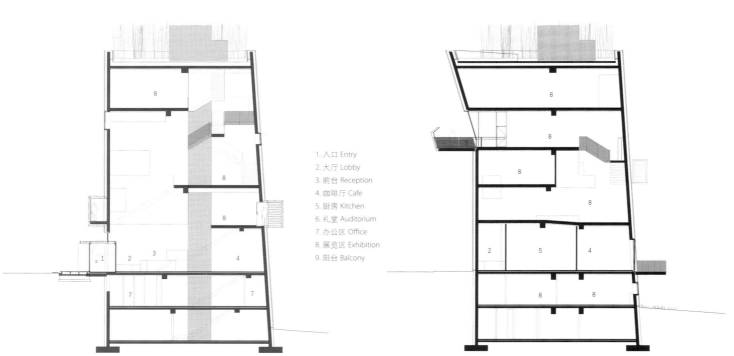

1. 入口 Entry
2. 大厅 Lobby
3. 前台 Reception
4. 咖啡厅 Cafe
5. 厨房 Kitchen
6. 礼堂 Auditorium
7. 办公区 Office
8. 展览区 Exhibition
9. 阳台 Balcony

Credits

Location: Hamarøy, Norway
Date: 1994 ~ 2009
Building Area: 2,508 m²
Architects: Steven Holl (design architect)
Noah Yaffe (associate in charge - construction documents)
Francesco Bartolozzi, Ebbie Wisecarver (project team - construction documents)
Peter Englaender, Chris McVoy (project team)
Erik Fenstad Langdalen (project architect - design development)
Gabriela Barman-Kraemer, Yoh Hanaoka, Justin Korhammer, Anna Müller, Audra Tuskes (project team - design development)
Mechanical Engineer: Ove Arup

Awards
AIA NY Honor Award
North Norwegian Architecture Prize
Photographers: Iwan Baan, Steven Holl Architects

蓝天下,绿地中,"玻璃巨石"散发出的灯光更显博物馆的弧形曲线美
The curved museum looks more elegant set off by soft lighting, green land and blue sky....

| Steven Holl Architects in collaboration with Solange Fabiao |

Cité de l'Océan et du Surf
城市海洋科学博物馆

2012 Gold Medal Recipient   2012金奖

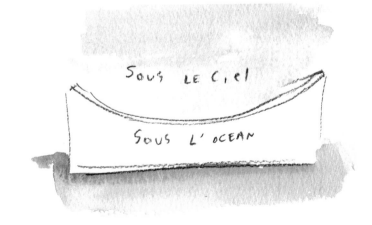

城市海洋科学博物馆是一座探索海洋以及海洋在人类生活、科学以及生态学中所扮演角色的博物馆。该设计由Steven Holl Architects 与Solange Fabião 联合完成，是2005年一项国际比赛的获奖作品，当时参赛的公司包括Enric Miralles/Benedetta Tagliabue, Brochet Lajus Pueyo, Bernard Tschumi 以及Jean-Michel Willmotte。

建筑的造型由立体概念"苍穹下/深海处"演化而来。内凹的"苍穹下"构建了主要的外部广场，"海洋广场"面向蓝天和大海，遥望远方的天际线。凸起的天花板结构构筑起了"深海处"的展示空间。

两块组成餐厅与冲浪亭的"玻璃巨石"使得户外中心广场显得更加灵动，与海滩上的两块巨石相呼应。建筑西南角是冲浪者聚会场所，上方是冲浪池，下方是门廊，连通着博物馆内的礼堂和展示厅。这个带遮棚的区域为室外互动，聚会以及大型活动提供了场所。

城市海洋科学博物馆的花园旨在将建筑融合于周围环境，让建筑与海平面相连。设计理念与地形恰到好处的融合使得建筑别具一格。公共广场采用葡萄牙鹅卵石铺设，覆盖有草坪以及天然植被。建筑内凹的广场向周边景色延伸而去。这些稍带有杯型边缘的花园将场所与当地的植被相融合，作为建筑的一种延伸，并且配合着博物馆的设施将用来举办节日或者日常活动。

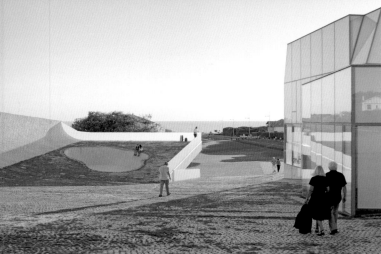

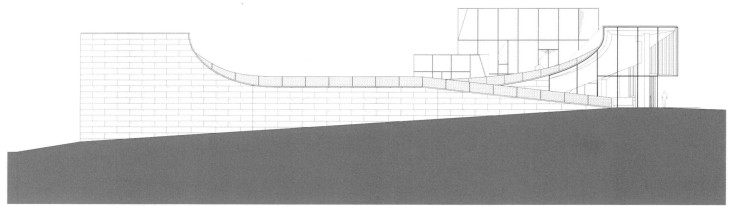

东立面
East elevation

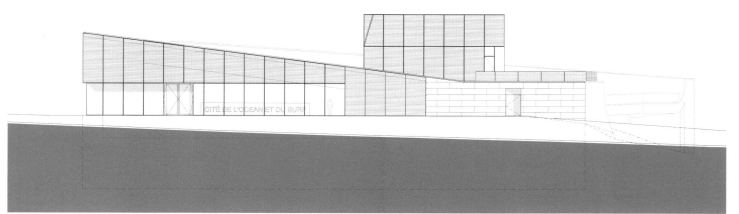

北立面
North elevation

户外广场铺制了精美的鹅卵石
The outdoor square is covered with delicate pebbles

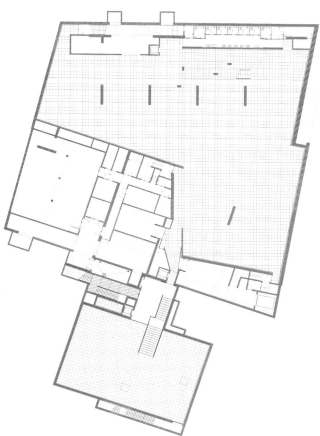

下凸的天花板形成深海般的感觉
A convex structural ceiling forms the "under the sea" exhibition spaces

The Cité de l'Océan et du Surf is a museum that explores both surf and sea and their role upon our leisure, science and ecology. The design by Steven Holl Architects in collaboration with Solange Fabião is the winning scheme from an international competition in 2005 that included the offices of Enric Miralles/Benedetta Tagliabue, Brochet Lajus Pueyo, Bernard Tschumi and Jean-Michel Willmotte.
The building form derives from the spatial concept "under the sky" / "under the sea". A concave "under the sky" shape forms the character of the main exterior plaza, the "Place de l'Océan", which is open to the sky and sea, with the horizon in the distance. A convex structural ceiling forms the "under the sea" exhibition spaces. Two "glass boulders", which contain the restaurant and the surfer's kiosk, activate the central outdoor plaza and connect analogically to the two great boulders on the beach in the distance. The building's southwest corner is dedicated to the surfers' hangout with a skate pool at the top and an open porch underneath that

内部楼梯
Interior staircase

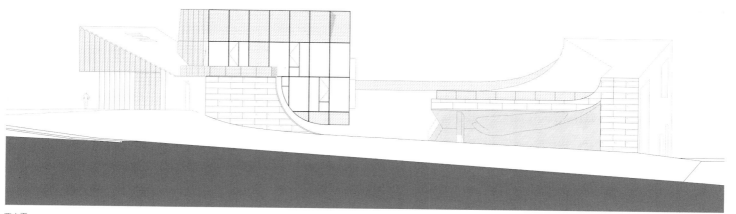

西立面
West elevation

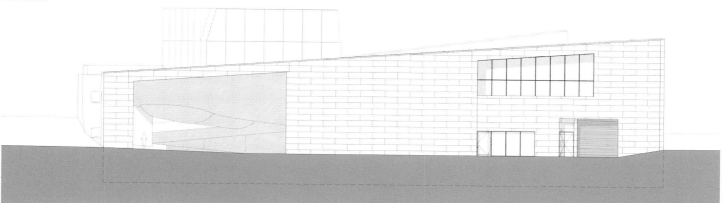

南立面
South elevation

两块"玻璃巨石"在灯光的照射下晶莹剔透
Two glass boulders shine with gorgeous lighting

connects to the auditorium and exhibition spaces inside the museum. This covered area provides a sheltered space for outdoor interaction, meetings and events.

The gardens of the Cité de l'Océan et du Surf aim at a fusion of architecture and landscape, and connect the project to the ocean horizon. The precise integration of concept and topography gives the building its unique profile. The materials of the public plaza are a progressive variation of Portuguese cobblestone paving with grass and natural vegetation. Towards the ocean, the concave form of the building plaza is extended through the landscape. With slightly cupped edges, these gardens, a mix of field and local vegetation, are a continuation of the building and will host festivals and daily events that are integrated with the museum facilities.

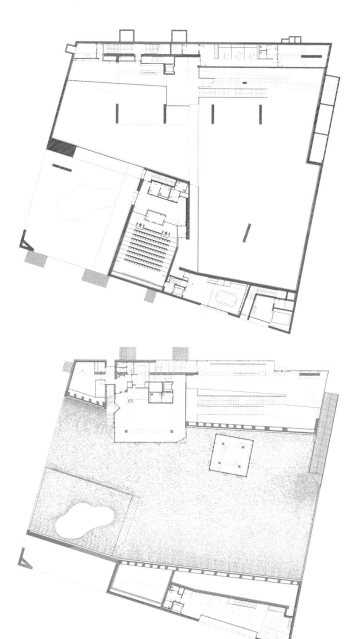

Credits

Location: Biarritz, France
Date: 2005 ~ 2011
Steven Holl Architects:
Design Architect: Solange Fabião, Steven Holl
Project Architect: Rodolfo Dias
Project Advisor: Chris McVoy
Assistant Project Architect: Filipe Taboada
Project Team: Francesco Bartolozzi, Christopher Brokaw, Cosimo Caggiula, Florence Guiraud, Richard Liu, Johanna Muszbek, Ernest Ng, Alessandro Orsini, Nelson Wilmotte, Ebbie Wisecarver, Lan Wu, Christina Yessios
Rüssli Architekten:
Project Team DD/CD: Justin Rüssli, Mimi Kueh, Stephan Bieri, Björn Zepnik
Photographers: Iwan Baan, Roland Halbe

| BIG-Bjarke Ingels Group |

8 House | 8字形大楼 |

2012 Institute Honor Awards for Architecture 　　2012建筑荣誉奖

评委评语

"8字形大楼"通过在这个混用、多户型住宅项目中加入一系列易入坡道技巧纯熟地再创造了水平社交连通性和社区街道的相互作用。这种巧妙的造型提供了一个活力充沛的雕塑外形,并创造了倾斜的"步行街"系统,给充满阳光和视野开阔的住宅单元提供了完全的深度。

人们真正"生活"在这个新创造的、集购物、餐饮、美术馆、办公设施、儿童托管、教育设施和孩子的嬉闹声于一体的社区里。这是一个综合、可借鉴的新类型。

Jury Comments

The 8 House masterfully recreates the horizontal social connectivity and interaction of the streets of a village neighborhood through a series of delightful accessible ramps in a mixed use, multifamily housing project. The skillful shaping of the mass of the facility provides an invigorating sculptural form while creating the ramped "pedestrian" street system and providing full depth dwelling units which are filled with light and views.

People really "live" in this newly created neighborhood with shopping, restaurants, an art gallery, office facilities, childcare, educational facilities and the sound of children playing. This is a complex and exemplary project of a new typology.

公寓楼外绿草如茵
The house backed by greenery

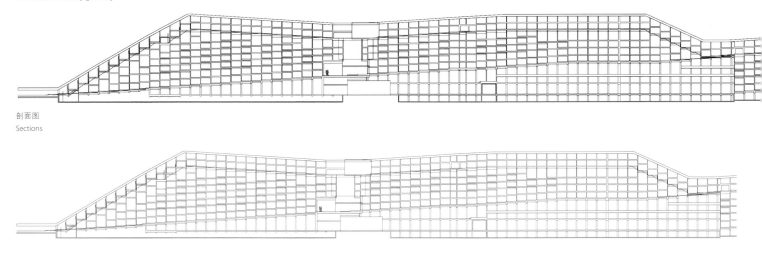

剖面图
Sections

大楼位于奥雷斯塔德南部哥本哈根运河边缘,在这里可以观赏到西阿迈厄岛的美丽景色。这是一栋真正意义上的大楼,为各个年龄层的居民提供了居住空间:年轻人和老年人,核心家庭和单身家庭,规模逐渐壮大或者变小的家庭。8字形大楼的中心被一分为二,形成两个不同的空间,大楼的中心主要是500平方米的公用设施。同时,这里还有一条9米宽的通道,连接两个周边城市空间:西部的公园和东部的通道区。商业空间和居住空间并没有分割成两个独立的结构体,而是分布在不同的楼层上。公寓被安置在顶部,商业空间在建筑的底部。因此,不同的楼层拥有其自身的特质:公寓层拥有良好的观景视角、阳光和新鲜空气,而办公室租赁区与街道上的生活相融合。

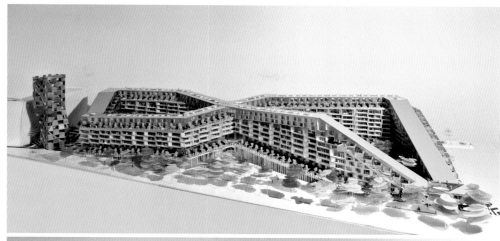

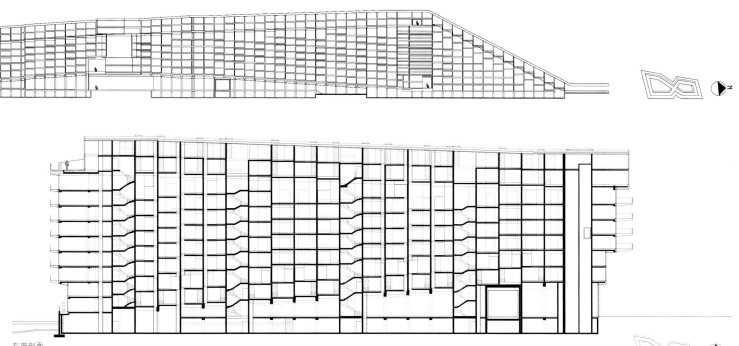

东南剖面
South-east section

夜晚,在灯光下熠熠生辉的公寓楼与水中的倒影相映生辉
At night, the light-lit building complements to its water reflection

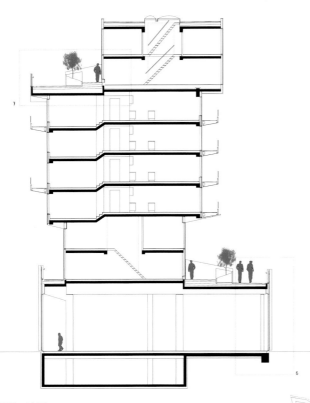

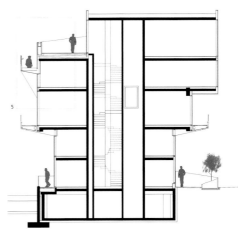

8 House is located in Southern Ørestad on the edge of the Copenhagen Canal and with a view of the open spaces of Kalvebod Fælled. It is a big house in the literal sense of the word. A house offering homes in all its bearings for people in all of life's stages: the young and the old, nuclear families and singles, families that grow and families that become smaller. The bow-shaped building creates two distinct spaces, separated by the center of the bow which hosts the communal facilities of 500 m². At the very same spot, the building is penetrated by a 9 meter wide passage that connects the two surrounding city spaces:

西北剖面及南剖面
North-west & south sections

远处一片绿草蓝天
Overlook the grass and the blue sky

滨水景观
Waterside view

灯光照耀下的大楼
Light lit house

远处一片绿草蓝天
Overlook the grass and the blue sky

供小孩玩耍的宽敞道路
The wide path as a playground for children

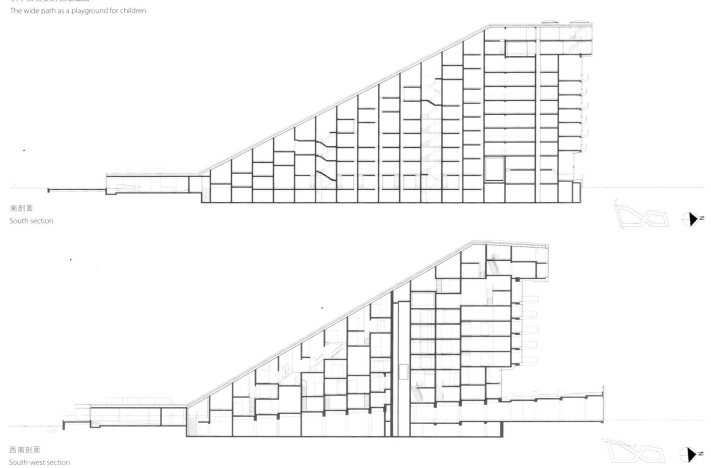

南剖面
South section

西南剖面
South-west section

the park area to the west and the channel area to the east. Instead of dividing the different functions of the building—for both habitation and trades—into separate blocks, the various functions have been spread out horizontally. The apartments are placed at the top while the commercial program unfolds at the

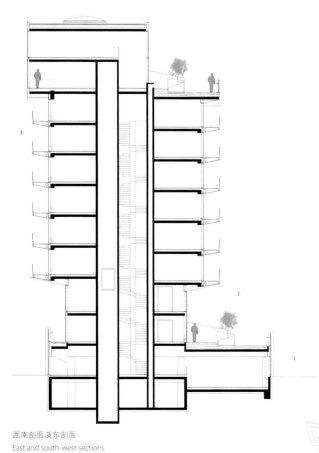
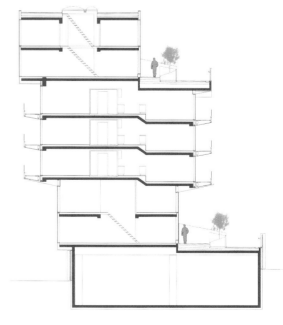

西南剖面及东剖面
East and south-west sections

楼梯与墙壁均被涂成白色
Staircase in the same white color with the wall

小孩的娱乐活动
Children's play

公寓内景
Views inside the apartment

公寓楼远景
Distant view of the house

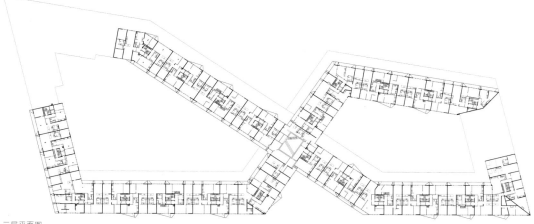

三层平面图
Apartment level 3 plan

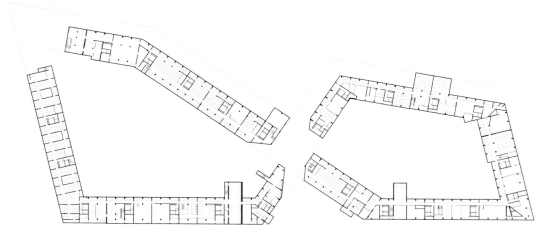

地下一层平面图
Basement level plan

Credits

Location: Copenhagen, Denmark
Area: 61,000 m²
Cost: EUR 92,000,000
Architects: BIG – Bjarke Ingels Group
Partner in Charge: Bjarke Ingels, Thomas Christoffersen
Project Leader: Ole Elkjaer-Larsen, Henrick Poulsen
Project Manager: Finn Nørkjær, Henrik Lund
Client: St. Frederikslund Holding
Collaboration: Hopfner Partners, Moe & Brodsgaard, Klar
Photographers: Jan Magasanik, Jens Lindhe, Jesper Ray, Ty Stange

远处一片绿草蓝天
Overlook the grass and the blue sky

滨水景观
Waterside view

灯光照耀下的大楼
Light lit house

供小孩玩耍的宽敞道路
The wide path as a playground for children

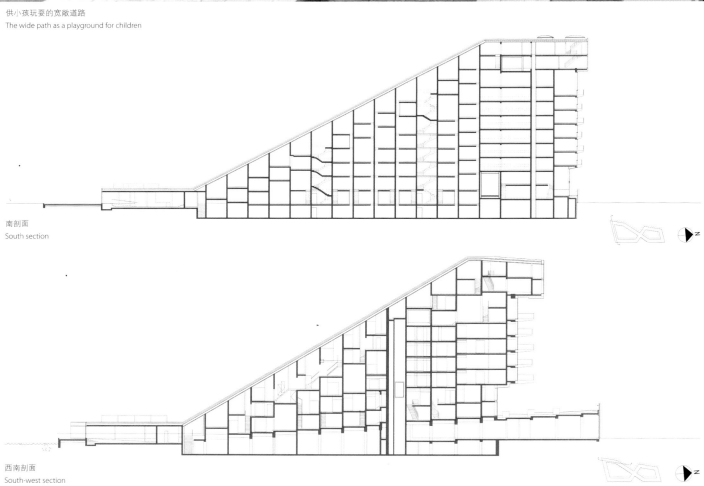

南剖面
South section

西南剖面
South-west section

立面详图
Facade detail

总平面图
Site plan

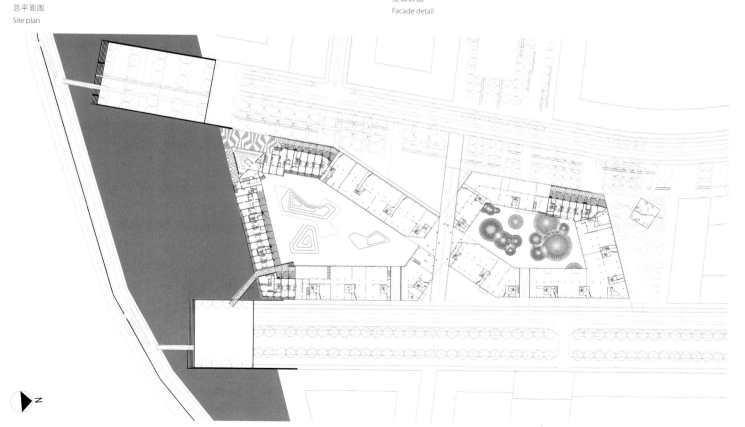

从下面的鲍什桥看盖茨中心东立面
View of Gates Center east elevation from below Pausch Bridge

| Mack Scogin Merrill Elam Architects |

Gates Center for Computer Science and Hillman Center for Future Generation Technologies
盖茨电脑科技中心和希尔曼未来发电技术中心

2012 Institute Honor Awards for Architecture　　2012建筑荣誉奖

评委评语

这个项目在一个城市大学和一个异常困难的基地里的比例十分完美。这个建筑不但和学校的文化与愿景相符，而且弥补了大学之前无疑缺失的连接部分。窗户和镀锌表皮令人惊讶地与校园组织完美融合而不显突兀。

或许该项目最美妙的部分要数由透明的室内玻璃窗、非反射的外部玻璃以及仔细放置的有角度的楼板创造的景观和视觉连接。

Jury Comments

This project is scaled perfectly within an urban campus and within a uniquely difficult site. The building not only matches the culture and aspirations of the school but also provides campus connections that had been clearly missing before. The fenestration and zinc exterior skin surprisingly relate beautifully to the campus fabric without being literal.

Perhaps the most wonderful aspect of the project is a set of views and visual connections created by transparent interior glazing, non-reflective exterior glazing as well as carefully placed and angled floor plates.

从下面的鲍什桥看盖茨中心东立面夜景
Gates Center east elevation viewed from Pausch Bridge at night

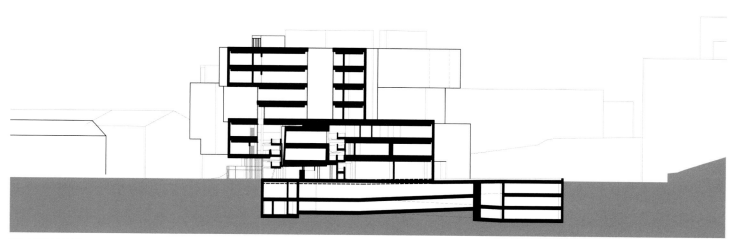

面向北面的东西剖面
East west section looking north

从盖茨中心上层向西望去
View to the west from upper level of Gates Center

盖茨电脑科技中心和希尔曼未来发电技术中心组成了卡内基梅隆大学西校区的电脑科学大厦。这座建筑包含这个电脑科学学院的三个部门，提供了办公室、会议室、开放协作空间、封闭室，还有一个阅览室连同更公众化的10个大学教室，一个250座的礼堂，一个咖啡厅和两个机房，可供120多个教师、350个研究生、100个研究员或博士后和50行政人员使用。

项目遵循了以下的设计理念：

一个学术机构的基本准则是允许释放个性。

聪明的、有创造性的人对自由选择的权利有一种本能的渴望。他们认为秩序高于程序，并要求在一个互相尊重的环境里保持个性。

建筑和景观建筑能容纳看起来不可调节的矛盾双方。

学术单元和系与系之间跨学科的协同作业已经成为卡内基梅隆标志性的优势之一。

由于电脑科学学院规模和多样性的增长，它的硬件必须支持和鼓励一个成功的集体差异环境。

盖茨中心的场地选择有一系列不寻常的技术冲突、功能和美学挑战，随着盖茨中心的加入，西校区在视觉上和硬件上都是名副其实的综合大学。

卡内基梅隆校园规划和许多它的建筑都有出众和悠久的特征，实现了学校创始人的最高理想。他们是那个时代的杰出工程，并成为了我们这个时代建筑和景观建筑的参考案例。

对于项目最好的诠释是：一个能凸显卡内基梅隆和电脑科学学院在世界级学术殿堂中的显著地位的建筑。

一个包容性的进程——强化大学内部互惠主义、巩固与周围社区的关系、深化自然资源管理，给项目增加了丰富性和可信度。

从希尔曼中心向南望去能看到远处盖茨中心的北立面
Looking south from Hillman Center to north facade of Gates Center beyond

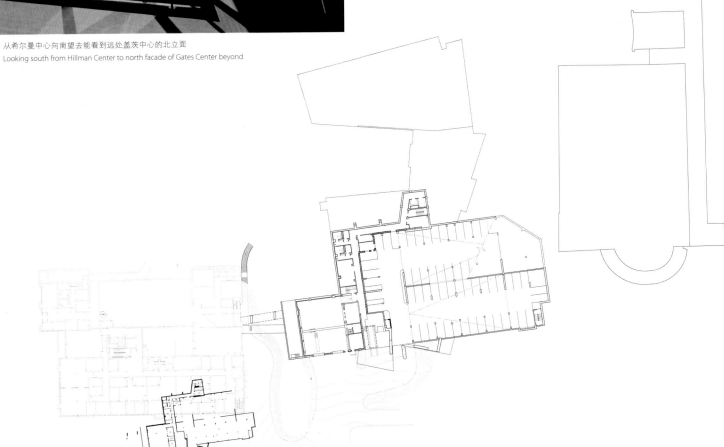

盖茨中心西立面
West elevation of Gates Center

从螺旋坡道望向下面的中庭和上面悬空的教室
Views from helix ramp to atrium below and suspended classroom above

技术实验室
Technical lab space

楼下的庭院
View to courtyard below

盖茨中心西北角
Northwest corner of Gates Center

六楼平面图
Level 06 floor plan

面向东面的南北剖面
North south section looking east

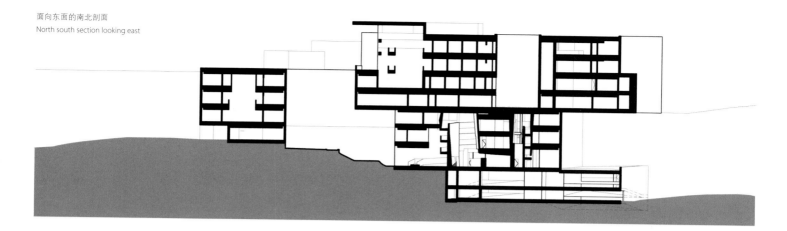

The Gates Center for Computer Science and the Hillman Center for Future Generation Technologies completes a computer science complex on Carnegie Mellon University's west campus. The building houses the three departments of the School of Computer Science providing offices, conference rooms, open collaborative spaces, closed project rooms and a reading room for more than one hundred and twenty faculty, three hundred and fifty graduate students, one hundred researchers or postdoctoral fellows and fifty administrative staff members along with a more public component of ten University classrooms, a two hundred and fifty seat auditorium, a café and two University computer clusters.

The following design principles have been used to guide the project conceptually:

The empowerment of the individual is fundamental to the mission of an academic institution.

Intelligent, creative people possess an innate desire for the freedom of choice. They privilege order over the systematic and demand the maintenance of individuality within a respected collective.

Architecture and Landscape Architecture have the capacity to sponsor the uncompromised coexistence of seemingly irreconcilable differences.

At Carnegie Mellon, interdisciplinary and collaborative work between both its academic units and its faculty has been and continues to be one of its defining strengths.

鲍什桥和远处的盖茨中心
View of Pausch Bridge and Gates Center beyond

盖茨中心西南角和前面闭合的桥接
Southwest corner of Gates Center with enclosed bridge connection in foreground

希尔曼中心和盖茨中心西北角
Northwest corner of Hillman Center and Gates Center

南视角
View from the south

As the School of Computer Science grows in size and diversity, its physical facilities must sustain and encourage a successful environment of collective difference.

The site chosen for the Gates Center has an unusually complex set of conflicting technical, functional and aesthetic challenges that, with the addition of the Gates Center, can serve to transform the West Campus area into a visually and physically integrated campus precinct.

The Carnegie Mellon Campus plan and many of its buildings have distinctive, enduring characteristics that embody the highest ideals of the institution's founders. They were exceptional works of their time and serve as exemplary benchmarks for an architecture and landscape architecture of our own time.

An architecture that represents Carnegie Mellon University's and the School of Computer Science's exceptional status among the world's leading academic institutions is best informed from within the project's own situation.

An inclusive process that reinforces reciprocity within the university, strengthens relations with its neighboring communities and deepens its commitment to the stewardship of our natural resources adds richness and credibility to this project.

Credits

Location: Pittsburgh, Pennsylvania, USA
Type: a school of computer science
Completion Date: fall 2009
Building Area: 19,324 m² and a 150 car parking garage
Architect: Mack Scogin Merrill Elam Architects
Landscape Architect: Michael Van Valkenburgh Associates
Local Architect: EDGE studio
Associate Architect: Gensler
Civil and Geotechnical Engineer: Civil and Environmental Consultants
Structural Engineer: Arup
Mechanical and Plumbing Engineer: Arup
Lighting Design: Arup
Electrical Engineer: Arup
Fire Protection and Life Safety Consultant: Arup
Communications and IT Consultant: Arup
LEED Consultant: Arup
Acoustical Engineer: Arup
Audiovisual Consultant: Arup
Security Consultant: Arup
Specifications Consulting: Collective Wisdom
Digital Assets Manager: CHBH
Cost Consultant: Heery International
Pausch Bridge Lighting Design: C & C Lighting
Parking Consultant: Tim Haahs
Hardware Consultant: Ingersoll Rand Security and Safety
Facade Assessment: Wiss, Janny, Elstner Associates
Construction Manager: P. J. Dick, Incorporated
Geotechnical Engineer: Construction Engineering Consultants
Surveyor: Gateway Engineers
Mack Scogin Merrill Elam Architects:
Principal: Mack Scogin, Merrill Elam
Senior Project Architect: Lloyd Bray
Senior Project Architect and Manager: Kimberly Shoemake-Medlock
Core Project Team: Alan Locke, Jared Serwer, Jason Hoeft, Clark Tate, Trey Lindsey, Jeff Collins
Project Team: B Vithayathawornwong, Dennis Sintic, Carrie Hunsicker, Misty Boykin, Barnum Tiller, Matt Weaver, John Trefry, Margaret Fletcher, Helen Han, Ben Arenberg, Brian Bell, Francesco Giacobello, Daniel Cashen, Janna Kauss, Patrick Jones, Cayce Bean, Jeff Kemp, Anja Turowski, Bo Roberts, Matthew Leach, Gary McGaha, Ted Paxton, Britney Bagby, Jacob Coburn, Amanda Crawley, Reed Simonds
Client: Carnegie Mellon University
Photographer: Timothy Hursley

鲍什桥中段看到的盖茨中心东立面和大型屋顶花园
Gates Center east elevation and intensive green roof viewed from Pausch Bridge at mid-span

| Andre Kikoski Architect |

The Wright at the Guggenheim Museum
古根海姆博物馆的莱特饭店

2012 Institute Honor Awards for Interior Architecture　　2012室内建筑荣誉奖

评委评语
这个项目处理谨慎，尊重建筑的原始本质。
在有限的空间和不宽裕的预算这些具有挑战性的条件下，这个项目完成得超乎寻常，特别值得一提的是程序的灵活性。
设计方法既节制又有趣，并且填补了博物馆整体运动和动态之间的细微差别。

Jury Comments
This project is sensitively handled and respectful of the essence of the original architecture. With the confined space and ostensibly modest budget, given those challenging constraints, this project is exceptional. Of special note is the programmatic flexibility. The design approach was controlled but playful, and complements the nuance of the museum overall movement and dynamic.

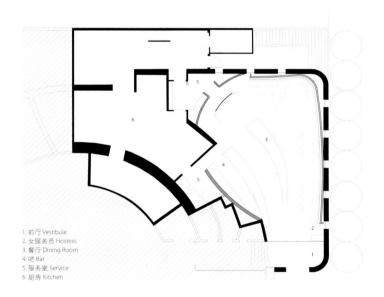

1. 前厅 Vestibule
2. 女服务员 Hostess
3. 餐厅 Dining Room
4. 吧 Bar
5. 服务室 Service
6. 厨房 Kitchen

在古根海姆博物馆内建造全新的饭店既是一种难以置信的荣耀，也是激动人心的挑战——这是大楼标志性的室内的第一处扩建。建筑师试图为大楼建造一个现代化且预算合理的149平方米的添加物。

设计方案参照了博物馆的设计，但绝非复制。设计过程中，建筑师将潜在的建筑几何加入到动态的空间效果里。雕塑的形态营造出了外翻的天花板。浪状的墙壁变成了舒适的座椅。弧形酒吧和公共餐桌使空间更具动感。玩味的造型使游客获得多样的体验。

该大楼由创新的、富有现代感的材料设计而成，极富质感。这其中包括了含有光纤层的胡桃木、创新性的定制金属制品的闪亮外壳、无缝可丽耐大理石表面、灰色材质的发光面板以及层叠的紧膜白色天棚。这些材料和色泽汇集一处，完美地实现了利亚姆·吉利克的设计。

建筑的表面和材质都极具动感，营造出的变幻美感在照明材料和屋内光源的映衬下更富生机。饭店为用餐者提供了高雅动感的环境，这样的氛围既源于博物馆，又胜于博物馆，也由此获得"博物馆瑰宝"的美誉。

It was both an incredible honor and an exhilarating challenge to create The Wright, the new restaurant at the Guggenheim Museum which is the first addition to the building's iconic interior. The architects sought to create a contemporary response to complement the building with a modest budget and 149 m² in which to work.

The design solution references the building's architecture without repeating it. In the process, underlying architectural geometries were transformed into dynamic spatial effects. The sculptural forms create a flared ceiling. The undulating walls become comfortable seating. The arced bar and communal table animates the space. The playfulness of these forms offers a dynamic experience for visitors. This project is highly tactile and crafted from innovative, contemporary materials. These include fiber-optic layered walnut, a shimmering skin of innovative custom metalwork, seamless Corian surfaces, illuminated planes of woven grey texture, and a glowing white canopy of layered taut membrane. Together these materials and colors form a perfect complement to the site-specific artwork by Liam Gillick.

The surfaces and textures embody movement, creating an ever-changing aesthetic that is enlivened with subtle layers of illumination and glowing tiers of light that envelope the room. The space achieves an elegant and dynamic setting for dining that both celebrates the museum and transcends it, and is often referred to as "a gem in the Guggenheim".

浪状的墙壁变成了舒适的座椅
The undulating walls become comfortable seating

雕塑的形态营造出了外翻的天花板
The sculptural forms create a flared ceiling

Credits

Location: New York, USA
Area: 149 m²
Date: 2009
Design Company: Andre Kikoski Architect
Photographer: Peter Aaron at ESTO

弧形酒吧和公共餐桌使空间更具动感
The arced bar and communal table animates the space

| 城外住宅区中心 | 社区停车场 | 多车道大街和市中心区 |
| Uptown Center | Neighborhood Parkway | Multiway Boulevard and Mid-town Center |

鼓励在电车走廊沿线开发约557公顷的交通导向型社区，为什么不充分利用当地各个领域
Incent the formatting of 60 million square feet in new growth toward transit-oriented neighbor
DNA of Fayetteville's successful downtown—which everyone loves—with the local conditions of

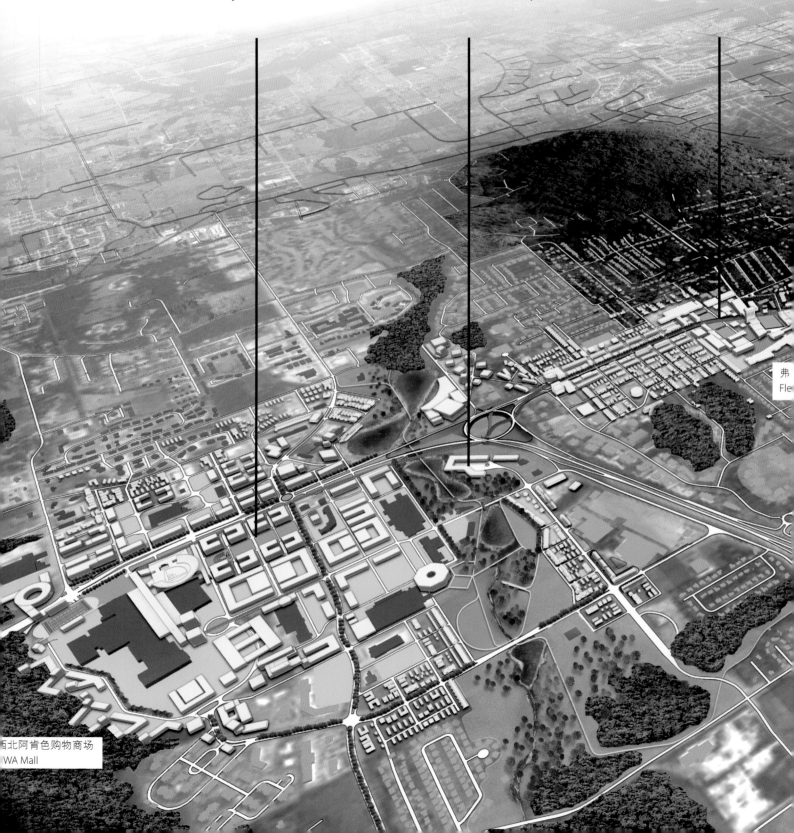

西北阿肯色购物商场
WA Mall

弗
Fle

| 社区林荫大道 | 阿肯色大学医学院和退伍军人管理局 | 商业区 | 阿肯色大学 |
| Neighborhood Boulevard | UAMS and Veterans Administration Campuses | Downtown | University of Arkansas |

…特维尔的商业区得到进一步发展？
…ment along the streetcar corridor. Why not breed the

| University of Arkansas Community Design Center |

Fayetteville 2030:
Transit City Scenario
费耶特维尔2030：交通之城

2012 Institute Honor Awards for Regional and Urban Design
2012区域和城市设计荣誉奖

评委评语

这个项目非常具有前瞻性。
项目的全面性和基础设施里公共空间特性的清晰可见性令人称道。
巧妙地保留了原来城镇的田园特征，同时加入了更多的现代化元素。

Jury Comments

The premise of this project is very forward-thinking.
There is a great appreciation for the comprehensive scope of this project as well as the clear visualization of the character of the public space within the scale of infrastructure.
The preservation of the rural character of the existing town as well as the addition of the more modern elements has been masterfully handled.

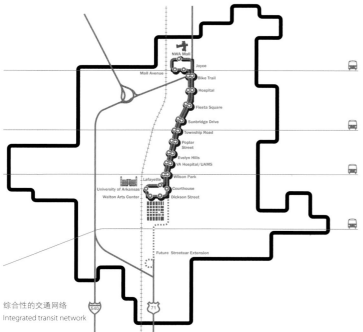

综合性的交通网络
Integrated transit network

现有的福布莱特高速公路交叉道和大型零售店
Existing Fulbright Expressway interchange and large-format retail

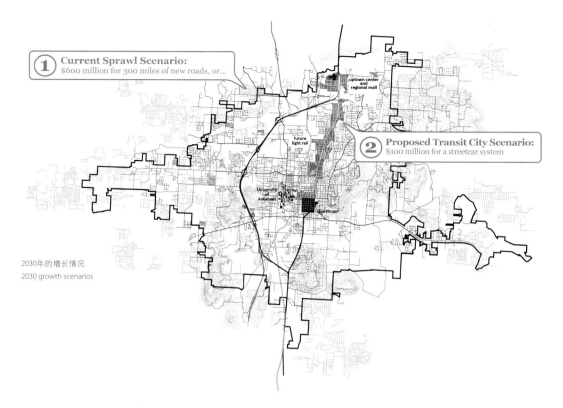

2030年的增长情况
2030 growth scenarios

机会

费耶特维尔2030规划项目旨在将阿肯色的人口在73 000的基础上增加55 000多人,住房量在32 000套的基础上增加28 000套。由于费耶特维尔目前有将近50%的建成环境难以保存到2030,这正为这座城市提供了一个机会来重新规划一个更智能的未来。规划的其中一个焦点是公共交通与土地利用的联系。试问,如果80%的开发都聚焦于学院大道(费耶特维尔的主要商业走廊)沿街近一千米的电车系统,情况会如何呢?

改造日渐衰退的商业走廊

像学院大道这样拥有多年历史,被零售店、大型停车场以及其他机动车设施占据的商业动脉曾经是强有力的经济发展工具。可惜,现在这些设施的经济利用价值已经消失殆尽,从那些毫无生机的购物商场、被遗弃的带状商业中心以及低出租率的沿街建筑就可以看出来。学院大道通向相对密集的一环郊区和市中心,是开发综合性城市社区的绝佳地带,同时也可以赋予新社区以费耶特维尔传统主街道的独特活力。该方案将学院大道公路改造成了一个多模式干道,包

现有的沃尔顿艺术中心
Existing Walton Arts Center

现有的城镇公路发展模式
Existing Township Road development patterns

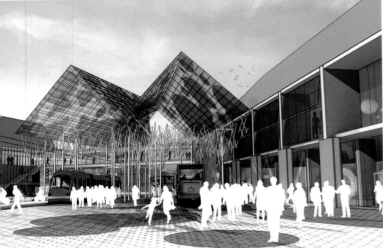

将学校、体育场与购物商场相结合
Integrate new high school and stadiums into shopping complex and transit galleria

将转车台与服务设施融入集市广场
Integrate streetcar turnaround barn, and service infrastructure into the market square

加强市中心与大学的联系
Intensify connection between downtown and the university

街区
Blocks

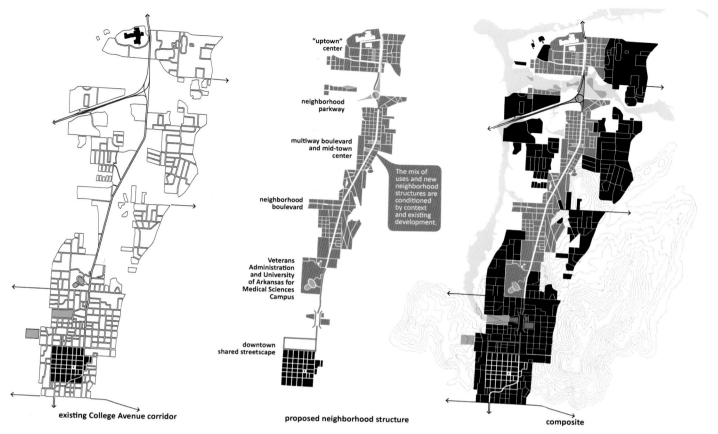

existing College Avenue corridor

"uptown" center
neighborhood parkway
multiway boulevard and mid-town center
neighborhood boulevard
Veterans Administration and University of Arkansas for Medical Sciences Campus
downtown shared streetscape

The mix of uses and new neighborhood structures are conditioned by context and existing development.

proposed neighborhood structure

composite

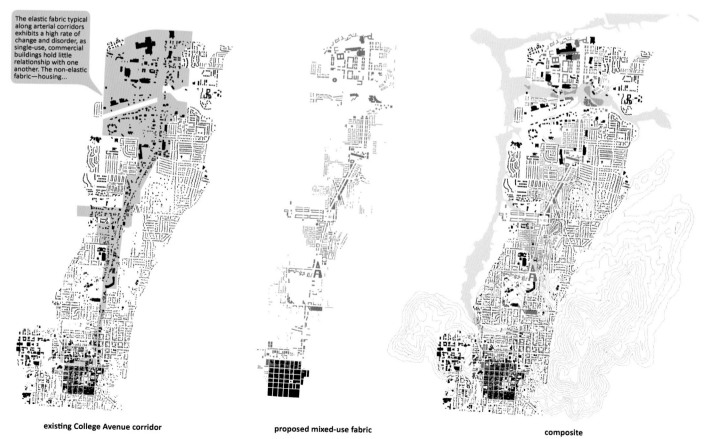

The elastic fabric typical along arterial corridors exhibits a high rate of change and disorder, as single-use, commercial buildings hold little relationship with one another. The non-elastic fabric—housing…

existing College Avenue corridor

proposed mixed-use fabric

composite

建筑
Buildings

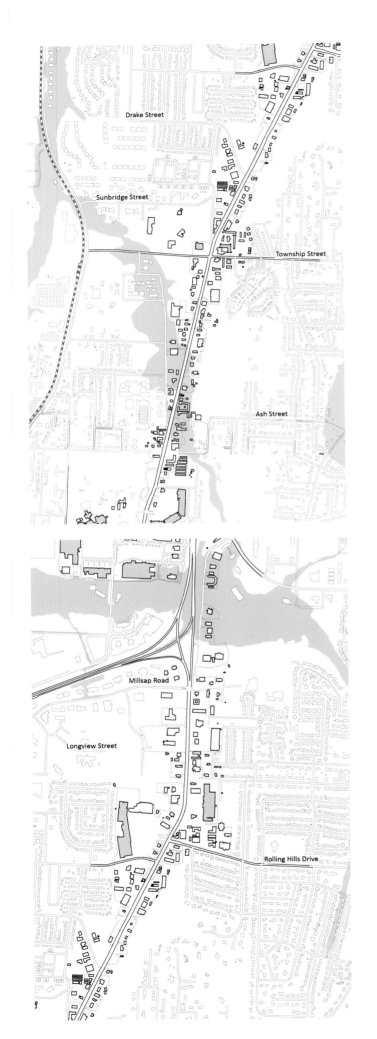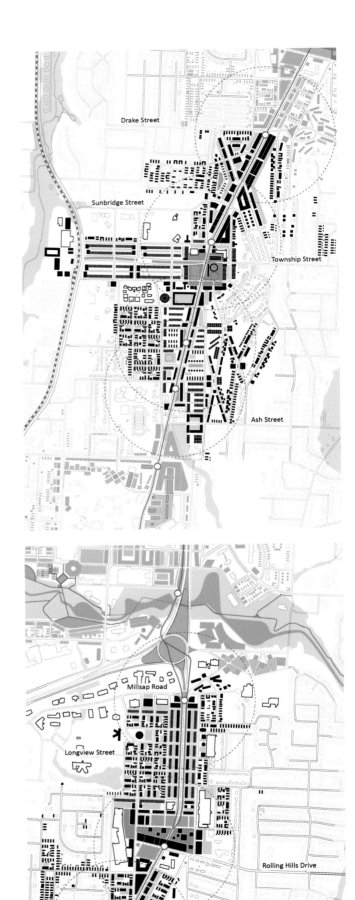

括人行道、自行车道、公交道等各种不同速度的交通模式。其目的在规划出具有明晰边界和中心的理想社区，以及相互连接的街道网络，同时为低密度（每英亩7~8个单元）和高密度（每英亩60~90个单元）城市区块服务。

交通之城——富足项目

典型的费耶特维尔家庭的交通成本是其年收入的29%，比19%的国家平均水平高出许多。2030交通之城方案是一个富足项目，因为在这个方案下，每家每户的交通成本将降到年收入的16%。

城外住宅区
Uptown neighborhood

立体交通道
Flyover traffic interchange

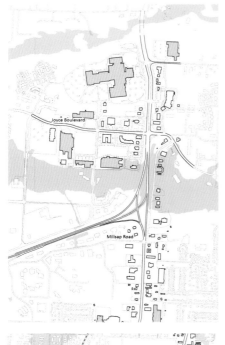
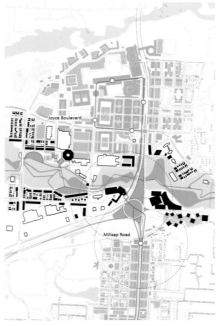
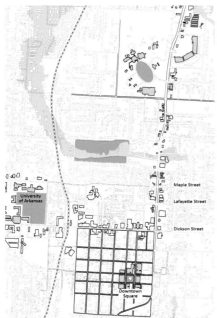
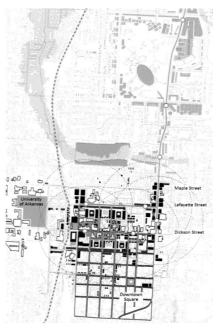
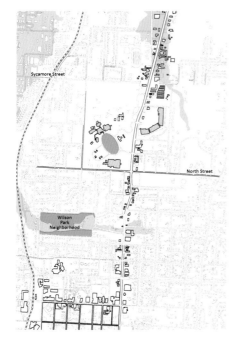
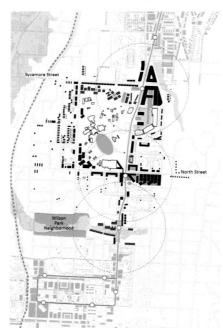

066 International New Architecture

新的之字形临山公路网络将现有的分离的住宅区连接起来
New hillside switchback fabric to link existing, but disconnected housing development

集合了电车站和新的雇佣中心的门户结构
Gateway structure housing transit stops and new employment center

The Opportunity

By 2030 Fayetteville, Arkansas is projected to add more than 55,000 people to its population of 73,000, entailing an additional 28,000 housing starts for a city with 32,000 dwelling units. Since close to 50% of Fayetteville's built environment projected to exist by 2030 has not yet been built, an opportunity exists to plan an even more intelligent future. One future scenario may focus on the connection between public transit and land use. We ask: what if 80% of the future growth (60 million square feet of conditioned space including 23,000 housing units) was incented to locate around a six-mile streetcar system proposed for College Avenue, Fayetteville's primary commercial corridor?

Retrofitting a Declining Commercial Corridor

Thousands of aging post-WWII commercial arterials like College Avenue—

创造可行的综合性城市脉络，将现有的开发项目纳入其中
Create walkable mixed-use urban fabrics that incorporate existing big box developments

dominated by strip retail, large parking lots, and other auto-oriented land uses—were once powerful economic development tools. Their economic usefulness has become exhausted, evidenced by dead malls, abandoned strip centers, and low-rent roadside structures cluttering large swaths of commercial arterials. College Avenue's passage through moderately dense first ring suburbs and downtown neighborhoods presents a ready opportunity to develop mixed-use walkable urban neighborhoods with a spatial vitality akin to Fayetteville's traditional main street. The project converts College Avenue highway into a multi-modal boulevard accommodating pedestrians, bicyclists, public transit, and traffic of various speeds. The goal is to deploy new development in imagable neighborhood configurations with clear edges and centers, and an interconnected street network serving both low (maintain current 7~8 units an acre) and high density fabrics (60~90 units an acre).

Transit City as a Prosperity-Building Program

A typical Fayetteville household spends 29% of its annual income on transportation, far above the national average of 19%. The 2030 Transit City Scenario plan is a prosperity-building program since the average household in rail transit cities spent 16% of its annual income on transportation.

改善有轨电车,将费耶特维尔的两个主要的历史住宅区衔接起来
Elevate streetcar and reconnect Fayetteville's two primary historic residential districts severed by the highway

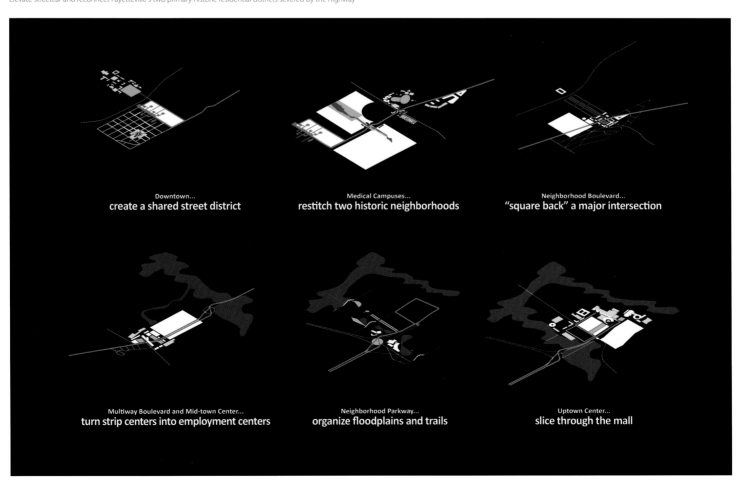

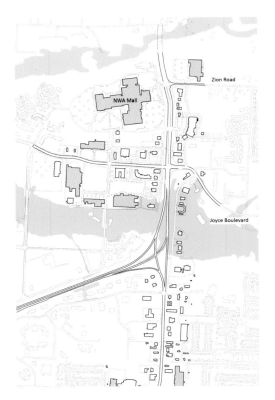

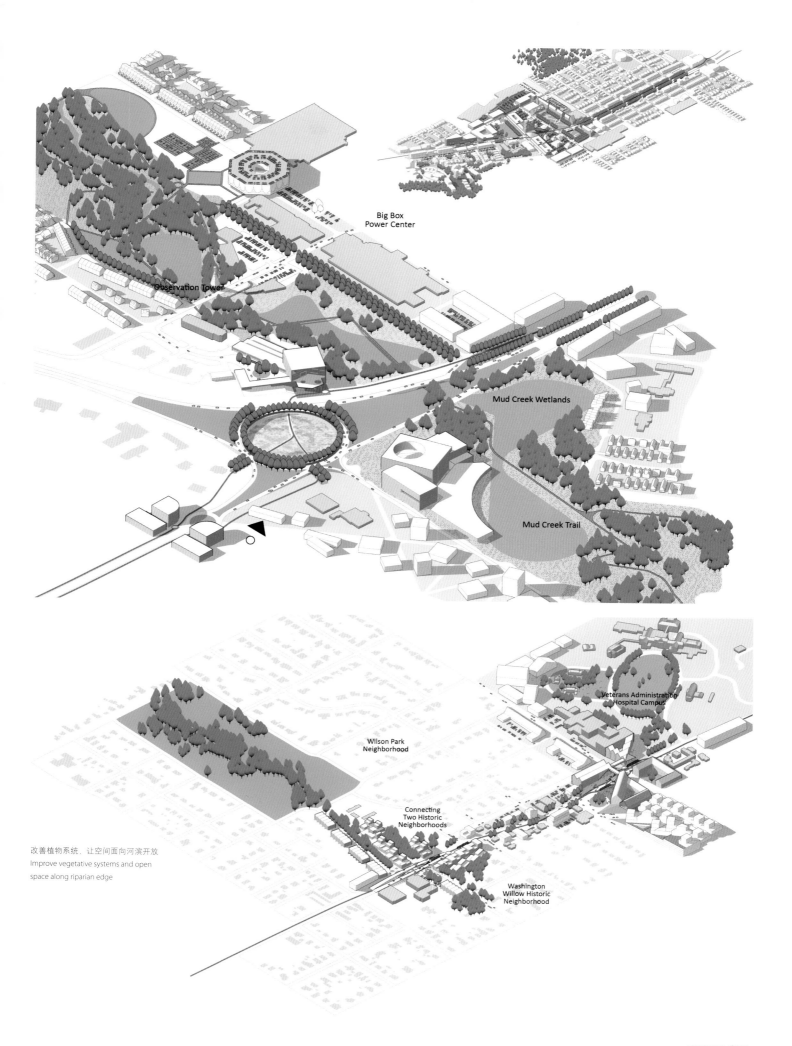

Grangegorman Master Plan

2012 Institute Honor Awards for Regional and Urban Design
2012区域和城市设计荣誉奖

Grangegorman总体规划

| Moore Ruble Yudell Architects & Planners |

项目最高目标之一便是将一个没有藩篱的基地重新带入都柏林浓密的城市脉络
One of the project's highest goals is the reintegration of a walled-off site back into the dense city fabric of Dublin

评委评语

这个项目展示了一个令人印象深刻的区域规划手段，有清晰系统的设计战略。对历史建筑的重新使用表现得非常成功。
新建筑和公共空间间隙里的透明感令人印象深刻。基本设计策略图示清晰有力。

Jury Comments

This project presents an impressively comprehensive approach to site planning, with clear and systematic design strategies. The adaptive reuse of the historic existing buildings appears to be very successful.
The sense of transparency both within the new buildings and through the interstitial public space is impressive. The diagrammatic representation of the fundamental design strategies was both clear and compelling.

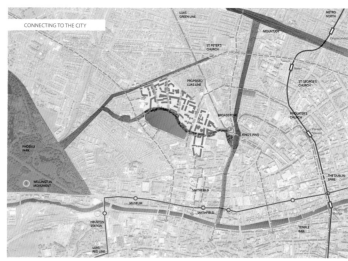

爱尔兰都柏林的Grangegorman小区总体规划项目给社区营造了热情的空间，同时与基地的过去保持关联，并成为了连接城市的枢纽
The Grangegorman Urban Quarter Master Plan in Dublin, Ireland creates inviting spaces for the community while maintaining connections to the site's past and linking together areas of the city

能够获得2012美国建筑师协会的国家荣誉奖对Grangegorman规划项目来说是一个崇高荣誉，因为这是美国建筑师协会在全球优秀项目中颁出的最高奖项。这个伟大的成就实现了我们对Grangegorman项目的最高期望和憧憬。

我们完全相信在这个充满挑战的年代对于爱尔兰年轻人的教育和未来投资的关键性，更希望这个享有声望的奖项能传播并在未来提升这个重要的Grangegorman项目的支持度和关注度。

我们对Grangegorman项目的目标是丰富都柏林市的城市脉络——使其成为一个充满生气和建筑价值的地方，不仅满足这个区域直接用户的需求，还能满足整个社区在21世纪及以后的需求。

詹姆士·奥康纳
美国建筑师协会会员
Moore Ruble Yudell Architects & Planners事务所负责人

Receiving the American Institute of Architects (AIA) 2012 National Honor Award is an extremely significant honor for the Grangegorman Master Plan, because it is the AIA's highest recognition of outstanding projects around the world. This is a tremendous achievement that represents a crucial validation of our best hopes and aspirations for the Grangegorman project.

We absolutely believe in the critical investment in the education and future of Ireland's young people during these challenging times, and are hopeful that this prestigious award will help to publicize and further increase the support and momentum for the important Grangegorman project.

Our goal for Grangegorman is to create a place which will enrich the city fabric of Dublin—a place full of vitality and architectural value to serve the needs of not only the immediate users of the quarter but indeed the entire community in the 21st century and beyond.

James Mary O'Connor, AIA
Principal
Moore Ruble Yudell Architects & Planners

未来的都柏林理工学院图书馆
Future DIT Library

未来的都柏林理工学院商学院大楼
Future DIT School of Business Complex

爱尔兰都柏林Grangegorman总体规划方案旨在为都柏林理工学院和爱尔兰国家医疗保健服务中心设计一个新的可持续性校园环境，并保留这里悠久的历史背景，加强这一地区同城市环境之间的联系。文化花园将校园中的旧式建筑和新式建筑有机衔接了起来。该设计将为校园提供世界一流的创新型便利设施，并通过现代手法将传统的学院建筑风格进行完美诠释。

一条贯穿东西方向的步行大道上融入了许多条绿化带。该规划在校园中设置了两个主要活动中心：一处是图书馆广场，这个广场面向西侧开放，是校园的"心脏"部位。另一处是艺术论坛，旁边设有剧院、博物馆和展览画廊。而一条"主路"将校园内各处重要地点连接了起来，主路旁边通往各个建筑的"小径"，绿草茵茵、生机盎然。一系列南北景观带呈径向延伸，与都柏林理工学院和爱尔兰国家医疗保健服务中心各入口相衔接。学生公寓和文化设施位于运动场边，从这里可以远眺市区景象和都柏林山脉景色。

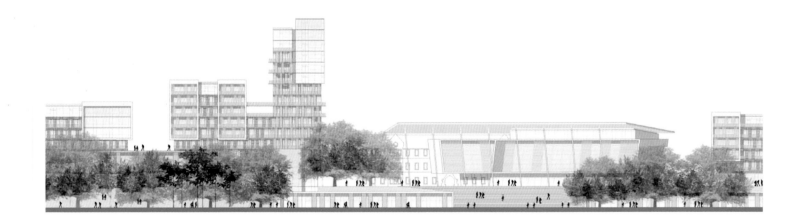

蛇形走道结合了运动设施和上面的学生公寓
The Serpentine Walk incorporates sports facilities below, with student residential housing above

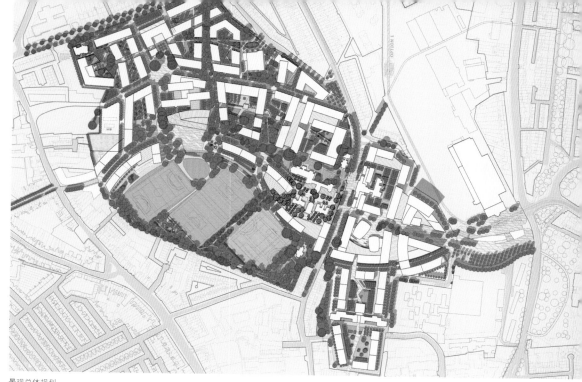

景观总体规划
Landscape Site Master Plan

未来的都柏林理工学院商学院大楼
Future DIT School of Business Complex

未来的都柏林理工学院商学院大楼
Future DIT School of Business Complex

未来的都柏林理工学院图书馆
Future DIT Library

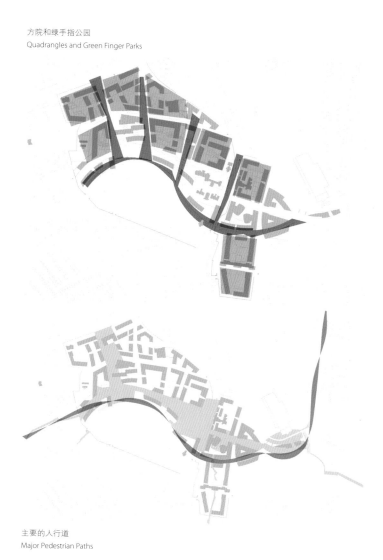

方院和绿手指公园
Quadrangles and Green Finger Parks

主要的人行道
Major Pedestrian Paths

都柏林理工学院中央图书馆可以欣赏到标志性的都柏林市景观和南部的山脉
The DIT Central Library provides iconic views out to the city of Dublin and the mountains to the south

一系列的方院和庭园被城市和景观小道连接,形成了一个四通八达的校园
A series of quadrangles and courtyards are linked by both urban and landscaped pathways to form a legible and interconnected campus

规划通过保护老树、开放空间和原有的历史建筑来维持与基地历史的联系
The Master Plan maintain links to the past history of the site by protecting mature trees, open space and existing historical buildings

"蛇形走道"提供了一条景观道，给学生公寓大厅带来了一条向南的绿边
The "Serpentine Walk" provides a landscaped path giving a south-facing green edge to student residence halls

The Grangegorman master plan creates a vibrant campus for the Dublin Institute of Technology (DIT) and Health Service Executive (HSE) by responding to the site's rich historical context and strengthening connections to the existing urban fabric. All of the standing "protected structures" within the site have been preserved and are connected to the new buildings through a Cultural Garden. The design offers world-class, innovative facilities for both DIT and HSE, enhancing their identity and image by employing a contemporary interpretation of traditional collegiate quads.

A major east-west pedestrian path seamlessly integrates with several significant green belts and circulation axes in the area. Within the campus, the master plan is focused on two centers of activity: Library Square, which serves as the "campus heart" toward the west, and the more public-oriented Arts Forum to the east, which is lined with theaters, museums and exhibition galleries. A formal "urban path" links the significant destinations on the campus while its counterpart—a "landscaped path"—provides a casual means of pedestrian circulation. A series of north-south landscape "fingers" extend radially to provide direct access to the separate DIT and HSE districts. Student housing and amenities are woven through the site along a sinuous landscape path at the edge of the sports pitches, looking out onto the city and Dublin Mountains beyond.

大学东面的主入口附近有一个重要的多模式交通枢纽,集中在原来的火车站——Broadstone终点站
The Master Plan features a major multi-modal transportation hub near the main eastern entry gateway to the campus, focused on a historic former train station, Broadstone terminal

两个核心
Two hearts

入口
Entry gates/Site connections

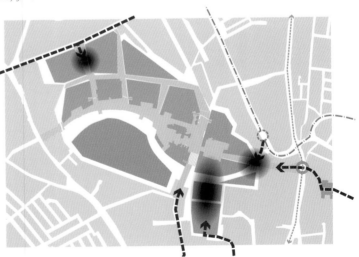

The DIT campus "Academic Heart" with the DIT Library on the right

Credits

Lead Firm, Design Architect: Moore Ruble Yudell Architects & Planners, Santa Monica
Principal-in-Charge: James Mary O'Connor
Partners: John Ruble, Buzz Yudell
Project Team: JT Theeuwes, Halil Dolan, Kaoru Orime, Nozomu Sugawara, Toru Narita, Tony Tran, Carissa Shrock, Matthew Henry, Tristan Hall, Joyce Ip Leus, Alon Averbuch, Simone Barth, Pooja Bhagat
Research & Marketing: Katie Carley
Graphic Design: Ken Kim
Models: Mark Grand, Alon Averbuch, Evan Henderson, Jenny Lee, Michael Dammeyer
Local Architects: DMOD Architects, Dublin
Director: John Mitchell. Ger Casey, Eoghan Garland
Architectural Conservation Consultant: Patrick Shaffrey Associates. Gráinne Shaffrey
Landscape Architect: Lützow 7. Jan Wehberg, Cornelia Müller, Tim Hagenhoff
Healthcare & Educational Environment Expertise: Prof. Bryan Lawson
Transport Planning/Civil & Infrastructure: Arup Consulting Engineers. Aidan Madden, Tiago Oliveira
Sustainability & Environmental Expertise: Battle McCarthy Ltd. Chris McCarthy, Neil Cogan
Digital Renderings: Shimahara Illustration
In-House Digital Rendering Team: Halil Dolan, Nozomu Sugawara, Matthew Henry, Tristan Hall
Watercolor Renderings: Tony Tran
Client: Grangegorman Development Agency
Chief Executive: Michael Hand, Gerry Murphy

面向街道的立面
Street elevation

分离的车库
Detached garage

| Studio H:T |

Shield House ｜ 金属盾包裹的房子 ｜

2012 Young Architects Award　2012青年建筑师奖

评委评语

布拉德就是这样一个人，坚定、聪明、富有洞察力，他致力于在教室里营造一种对话，塑造学生在工作室里有意识地提升探索新知的行为、构建概念框架，以及发展可持续设计策略。我们很欣赏他对未来建筑师的深远影响和模范作用。

Jury Comments

Brad is the real deal—committed, intelligent and insightful. He believes in creating a dialogue in the classroom, shaping how students participate in consciousness raising activities in the studio exploring new knowledge, building conceptual frameworks, and developing sustainable design strategies. We appreciate his perceptive and profound to instruction, engaging our future architects.

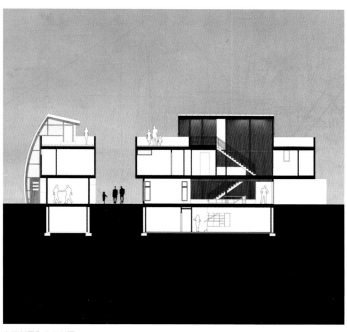

东西剖面和南北剖面
Sections—east/west & north/south

概念草图
Concept sketches

屋顶平台
Roof decks

入口
Shield entry

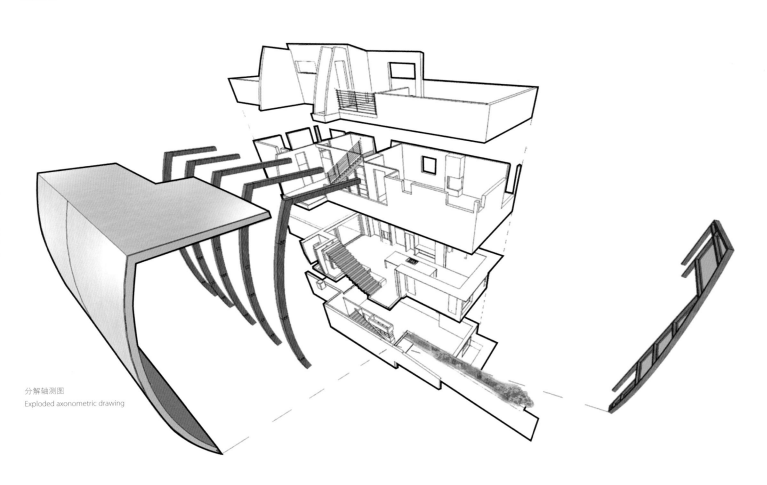

分解轴测图
Exploded axonometric drawing

重木框细部
Heavy timber frame detail

后院和分离的车库
Rear yard with detached garage

该项目将一个高高的、弯曲细长的流通空间与矩形的居住空间并置。建造这面高而弯曲的金属墙是由于受到了建筑体积的制约，同时可以保持对邻近的三层建筑的公共平台的私密性。这个元素无论在视觉和体验上都是住宅的焦点。它扮演了阳光接收器的角色，将早晨到午后的阳光带入住宅。夜晚，它对游客来说是一个发光而热情的帆。

This urban infill project juxtaposes a tall, slender curved circulation space against a rectangular living space. The tall curved metal wall was a result of bulk plane restrictions and the need to provide privacy from the public decks of the adjacent three story triplex. This element becomes the focus of the residence both visually and experientially. It acts as sun catcher that brings light down through the house from morning until early afternoon. At night it becomes a glowing, welcoming sail for visitors.

一楼的厨房和客厅
Main level kitchen / Living

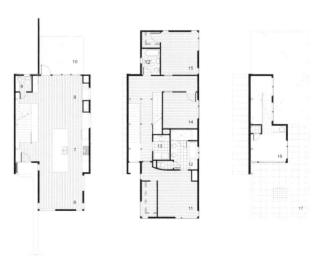

Credits

Location: Denver, Colorado, USA
Completion Date: 2010
Building Area: 302 m²
Architects: Studio H:T
Photographer: Raul Garcia

夜晚的外部平台
Exterior deck at night

住宅室内
Interior of shield

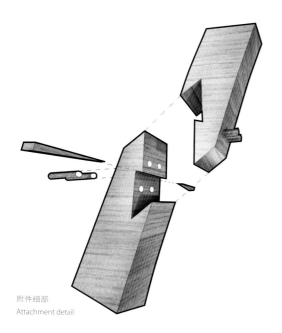

附件细部
Attachment detail

屋顶露台及花园
Roof deck garden

| Capita Symonds |

St Silas Primary School
圣西拉斯小学

剖面图
Section

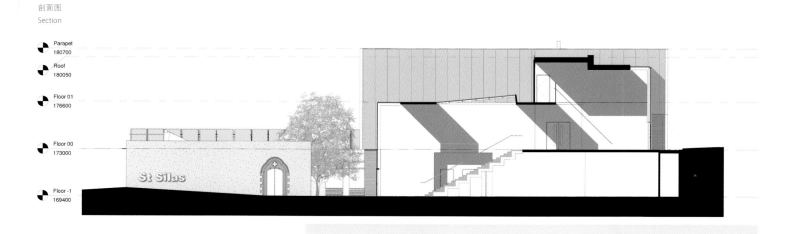

公共广场
Public square

立面图
Elevations

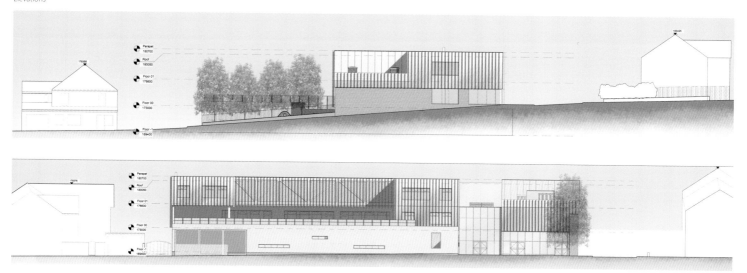

糖果色肋板
Candy colored fins

远眺奔宁山
View to Pennines

立面图
Elevations

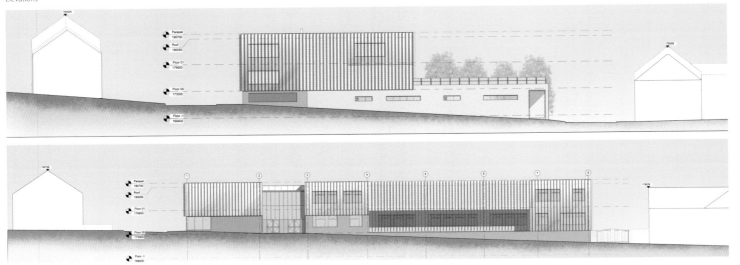

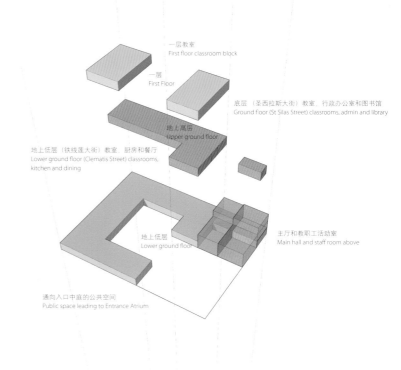

Capita Symonds建筑事务所代表布莱克本-达温委员会与当地教育企业密切合作，为这家拥有420位学生的小学提供了设计咨询服务——包括建筑设计、结构工程、机械和电子工程、景观设计、成本咨询、声学设计、项目管理、消防工程、规划以及交通规划。

如今，这所三层楼高的学校已对公众开放，它建于一处地势平缓的斜坡上，占地2 400平方米，从2010年夏季开始动工，仅用八周时间就完工，施工过程中采用多学科集成的BIM技术（建筑信息模型）。

该学校坐落在一群维多利亚风格排房中。项目所在地位于布莱克本中心西北1.6千米处，地处印度和巴基斯坦移民聚居区中心。这所英格兰教会经营的学校却有99%的生源具有伊斯兰教背景，堪称跨社区交流和社会融合的有趣案例。

Capita Symonds建筑事务所运用BIM技术加速了整个设计过程，将手工绘画理念融入三维建筑模型，使学校和当地社区都成为了社区交流的一部分，最终在几周之内实现了这一设计目标。

富有层次感的透明立面
Layered transparency

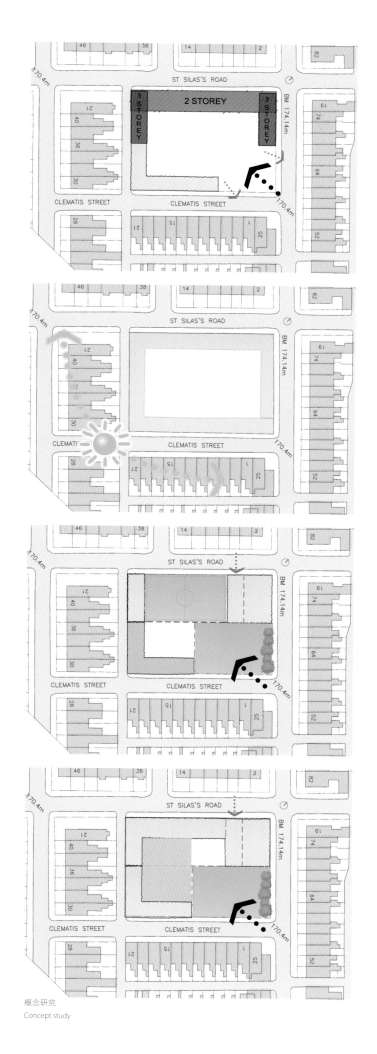

概念研究
Concept study

该事务所与当地教育机构"SHINE"及其合伙人Balfour Beatty一起合作，确保了项目的顺利实施，令老校区的拆迁在规划方案确定几天后就得以进行。

项目设计的主要挑战便在于既有场地的大小，长60米、宽40米，大大小于BB99建筑场地补贴的推荐面积。为了使所有教室能直通外面的游乐场，设计团队按照场地地形规划，充分利用地上地下空间，打造出了一座2 200平方米的三层建筑。通过横向的延伸让屋顶成为娱乐平台，同时用楼层的变化，使建筑设计简约大气。

学校由四幢相连的大楼构成，大楼围着中间的游乐场；其中一座单层大楼的顶部被设计成供人放松的平台，通过滑道与底楼相通，两栋三层大楼之间有通道相连，过道两侧都是教室，而楼上则有一间非正式的足球场；最后一栋便是容纳员工宿舍的学校主大楼，屋顶覆有人工植被。纵观整个设计，位于高层处的游乐场占地800多平方米，比之前既有的地面面积多出400多平方米。

各个大楼相互连接，通过合理配置使教学空间最大化，并使不同的学习风格融进多变的室内外教学区。教室围绕中心游乐场分布，越往上年级就越高，顶楼便是最高年级的教室。各大楼之间通行方便，大大便利了全校的活动和社会项目，同时图书馆通往大楼入口大厅，中间一条穿过大厅的过道，大厅一侧的窗户面朝大街，这样的设计又方便了社区和家长。学校中最关键的内部建筑——通透的门厅——焦点在于如同瀑布般高挂的台阶，那里作为中心聚会的场地，为学生们提供了晨间的休息区，也可用于学习，甚至还可以作为小型的节目表演平台。

学校大楼内，光线和色彩互相交织，五光十色的有机玻璃使用餐区内充满了

操场
Playdecks

彩虹般的光泽。大楼墙面外包裹的镀层是色彩鲜艳、光滑透明、质地坚固的有机玻璃肋片,这样的设计贵在有效地控制了成本,并将其他低成本的镀层材料包裹在内。学校不仅如同一个整体,而且还打破了传统的外立面观感,使观者在走近大楼时都能感受到透明和照明的交织。

建筑的有机玻璃取材于当地(从达温当地街区几千米外的地方),建筑立面反射出当地的景观,而形态、颜色和透明的"层积感"为这特别的场合营造出独特的学习环境。

庭院
Courtyard

从新滨大道看学校
View from New Bank Road

剖面图
Sections

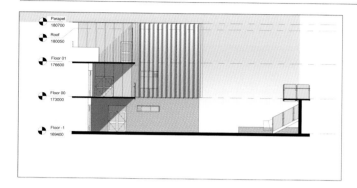

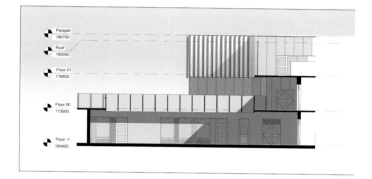

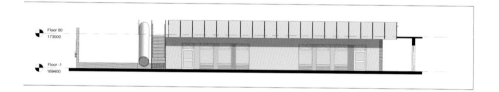

Working on behalf of Blackburn with Darwen Borough Council, in partnership with the local LEP, Capita Symonds provided a full range of multidisciplinary services—including architecture, structural engineering, mechanical and electrical engineering, landscape architecture, cost consultancy, acoustics, project management, fire engineering, planning and transport planning—for the two form entry, 420 pupil school.

The recently opened three story building was designed on a small, sloping, urban 2,400 m^2 site in just eight weeks in the summer of 2010 using BIM as a full multidisciplinary tool.

Located in a dense grid of Victorian workers' terraces, one mile northwest of Blackburn town center, the site is in the heart of a thriving community which has a large proportion of Indian and Pakistani residents. With 99% of pupils hailing from an Islamic background, this Church of England-run school provides an interesting example of cross community engagement and social mixing.

BIM techniques were utilized by the Capita Symonds team to accelerate the design process, enabling hand drawn concepts to be worked into three dimensions which were shown to the school and local community as part of the engagement process, with a final scheme reached in just a few weeks.

Working directly with the Local Education Partnership—SHINE—and its contactor partner Balfour Beatty, also meant cost and program certainty could be insured allowing demolition of the existing school to be undertaken just days after planning consent was obtained.

The primary challenge of the scheme design was the size of the existing site which, at just 60 meters by 40 meters, was considerably less than the typical

有机玻璃肋片
Perspex fins

recommended BB99 site allowance. In a bid to provide all classrooms with direct access to external play space, the team utilized the area's topography, resulting in a three story solution totaling 2,200 m² which is layered over and pushed into the site and terraced to maximize the external area with rooftop play spaces, while using level changes advantageously to create a simple, inspiring solution. The project comprises four linked blocks wrapping around a secure play courtyard: a single story block with a rooftop play deck linked to the ground with a tube slide; two three-story blocks linked by a bridge of class spaces of which the upper floor houses a mini-football pitch; and finally a main hall block with staff accommodation on top disguising plant areas. This design allows for over 800 m² of useable play space off the ground level, over 400 m² more than was previously provided on the existing flat site.

The disposition of the interlocking blocks is laid out to maximize teaching spaces and allow different learning styles with flexible indoor and outdoor teaching areas. The year groups spiral up in plan around the courtyard with the eldest at the top of the school. The flow between these blocks allows flexibility for whole school activity and community events while the library is accessible from the main foyer and acts as a bridge through the hall with a large window to the street, again encouraging community and parental use. The pivotal internal space in the school—a "through" entrance hall—is focused on a cascade of giant steps, acting as a central gathering space to provide a stopping point in the mornings, a special space for learning, or even a small performance venue. The play of light and color is deliberate throughout the school, with colored perspex step in-fills flooding dining areas with a rainbow of light. The "wrapping" elevation cladding system is a series of colored, translucent and solid perspex fins designed to create a cost effective rapid solution to enclose the otherwise relatively cheap envelope. This allows the building to appear as a whole mass, but also breaks up the facades as the viewer moves past the building with the whole exuding a playful mix of transparency and lightness.

The locally sourced perspex (from Darwen, just a few miles away within the borough) creates quick and "cropped" reflections of the local context when viewed in passing and adds to the "layering" of form, color and transparency to create a unique learning environment in a very special place.

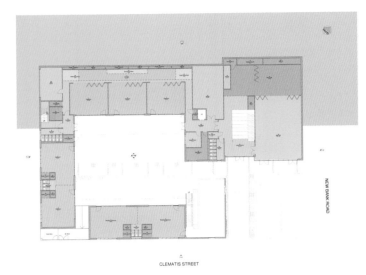

地上低层总平面布置图
Lower ground floor general arrangement plan

地上高层总平面布置图
Upper ground floor general arrangement plan

一层总平面布置图
First floor general arrangement plan

入口
Entrance space

餐厅
Dining hall

教室
Classroom

Credits

Location: Blackburn, UK
Start/Completion Date: 2010~2011
Lead Architect: Altaf Master
Design Director: Christopher Boyce
Team: Duncan Hammond, Dale Thomas and Tim Potter
Client: Blackburn with Darwen Borough Council
Photographer: Nick Guttridge

西侧外观
Exterior of the west part

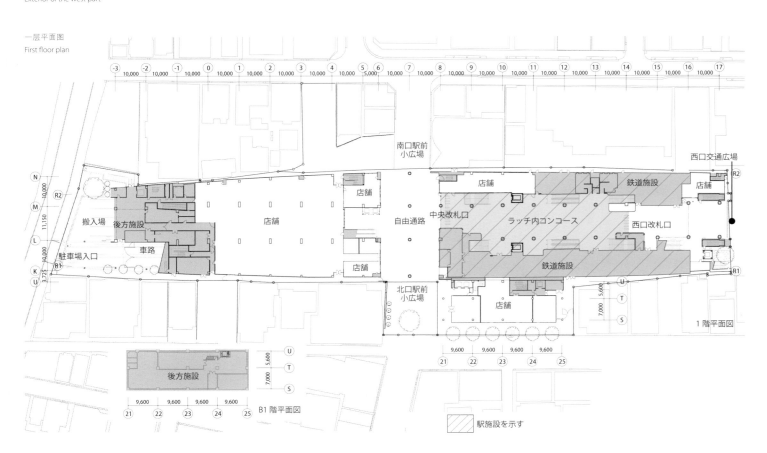

一层平面图
First floor plan

| SAKAKURA ASSOCIATES |

SEIJO CORTY
成城庭园

"SEIJO CORTY"是在小田急电铁"成城学园前"车站上面扩建的一个4层建筑，是一个由36家店铺构成的地下商业设施，也就是我们通常所说的"车站大楼"。提到车站大楼，大部分人都会认为那是一个位于车站上面的只谋求商业利益的"箱型商业大楼"，引导从检票口出来的乘客乘坐自动扶梯去商店或餐厅消费。与其相比，我们认为作为街道的中心设施，重点是为街道增添新意，将自然、街道、车站、商业融为一体，找回都市原本的魅力，营造出一个充满阳光、清风、绿意的空间。此外，由于设施对市区开放，也要展现出车站与商业相结合的繁华景象。设施名称"CORTY"（街上的庭园）一词是来源于意大利语"Cortile"，含义是庭园，这也是本次设计构思灵感的来源。

公共广场是该设施的中心，作为一个室外空间，它采用自然光，也无需使用空调，因此与室内相比，可削减75%的CO_2排放。此外，也可以让人们体验到通过四季变换而形成的"生态"空间。所谓的"生态"，并非是强迫人们忍耐酷暑严寒，而是让人感受到一个舒适的环境。如果我们每一个人都有意识地去改变，那么我们整个社会才能实现大量减少碳排放的目标。

二层平面图
Second floor plan

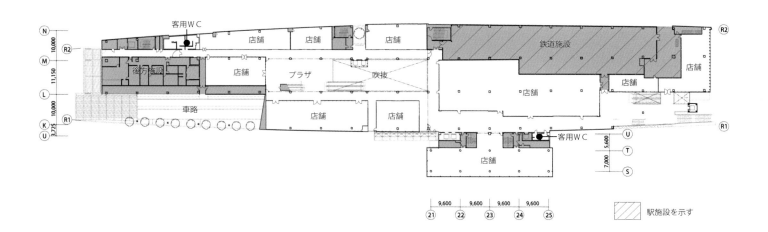

剖面图细节
Section detail

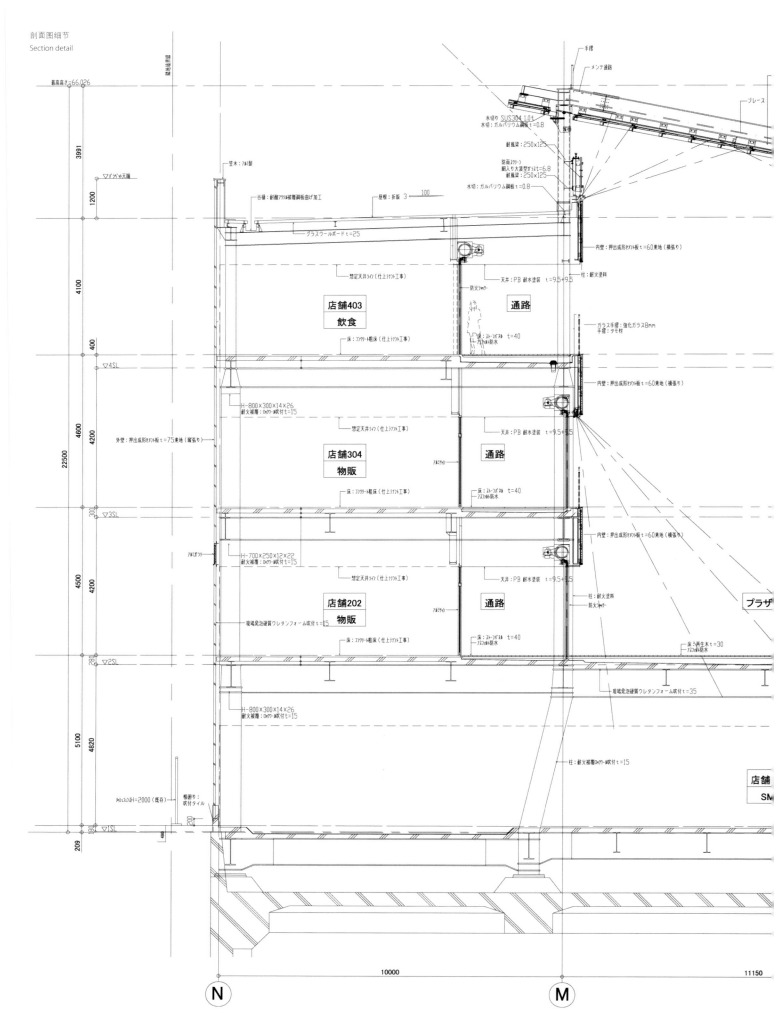

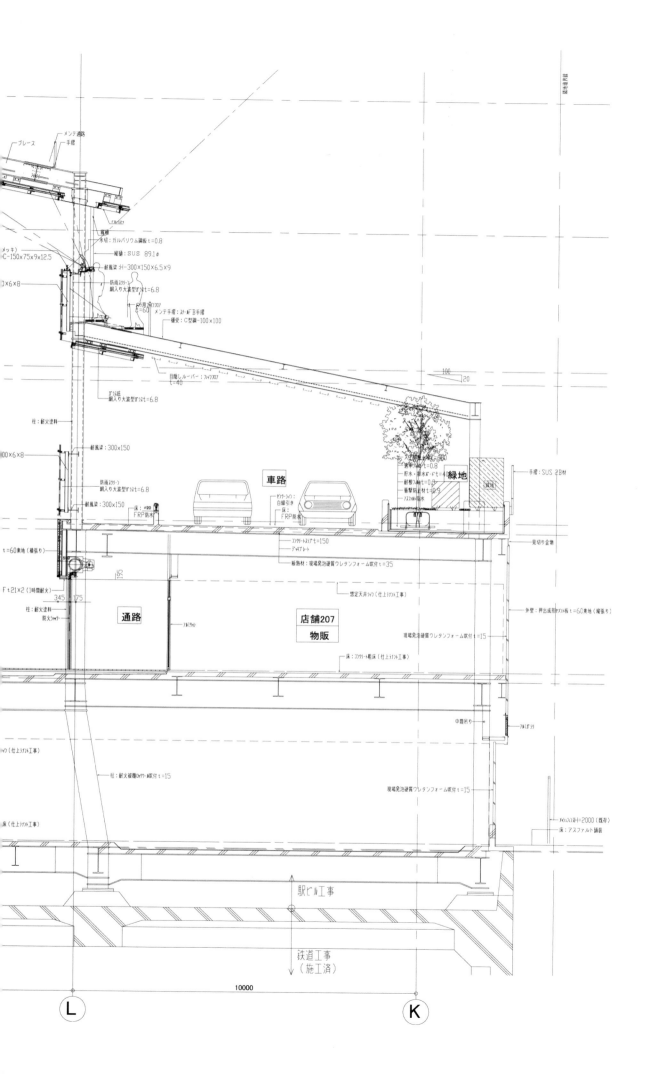

从东侧屋顶花园看公共广场
View of plaza from the roof garden of the east

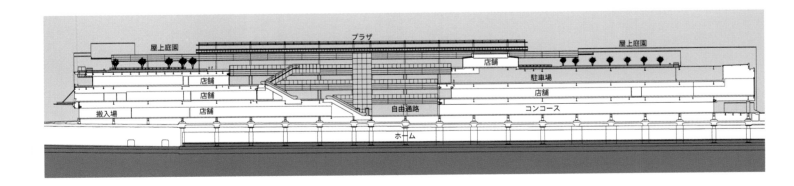

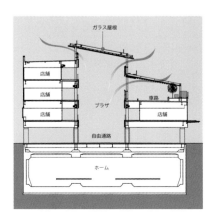

SEIJO CORTY is a four-story commercial facility comprising 36 stores to be built in an underground station under the city's continuous grade separation project. We are tired of looking at a station building pursuing its own interests and asserting its presence in the center of a city. Why not open an air hole in a city? Our idea is that it is more important to revive the city's original dynamism by linking nature, the city, the station, and commerce together and creating a transparent space filled with light, wind, and greenery. The name of the facility, CORY (a city yard), is derived from this idea.

The plaza in the center of the facility is designed to drastically reduce CO_2 (actually by 75%) by letting in natural light and making it an outdoor space without air conditioning. However, the significance of the plaza lies not in the figure, but in the opportunity that it offers for many people, including more than 80,000 train users, to experience an ecological space throughout the year. Ecology is not something that forces people to put up with the heat or the cold, but something that provides true comfort. If each person can truly realize this fact, then the high reduction target set for society at large can be achieved.

公共广场第3层
View of the 3rd floor of the plaza

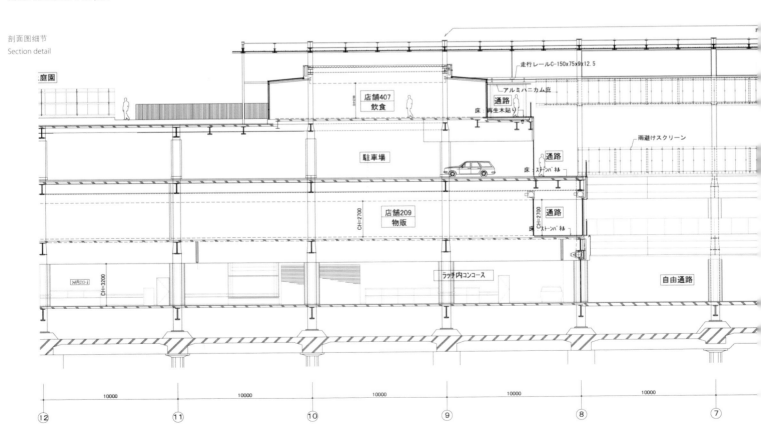

剖面图细节
Section detail

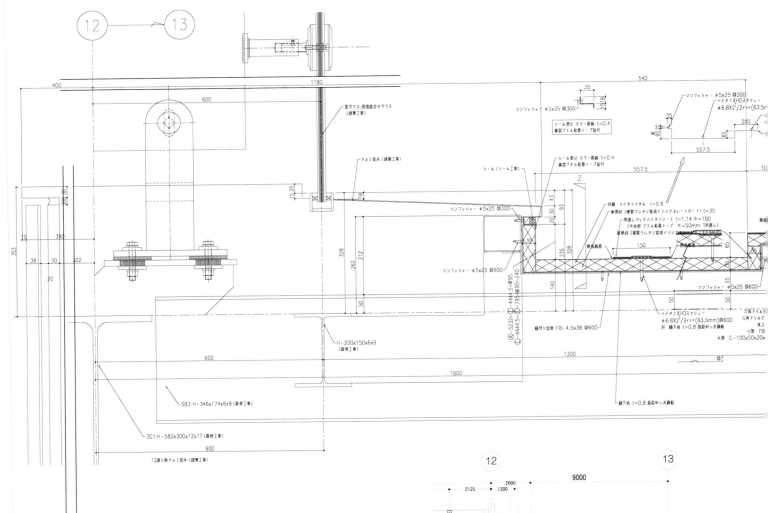

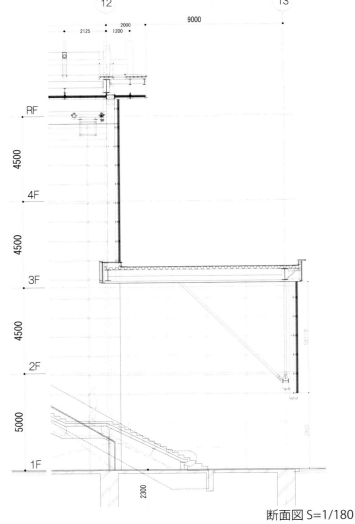

断面图 S=1/180

从"新宿"站到小田急电铁的"经堂"站有8千米,"经堂庭园"是在经堂站前建成的一个四层商业设施,共有47家店铺。它摆脱了传统的"箱型商业大楼"模式,将自然、街道、铁路、商业融于一体,目标是成为"都市核心"设施。

为实现这一目标,如果将大规模的公共空间作为庭园等使用的话,在企业收支上是不可行的。此外,"建筑基准法"中关于确保商业设施紧急避难楼梯的合理宽度及"避难安全检证法"中所规定的未来承租者进行店铺转租时,对天花板的高度要求及装饰材料相关的技术检定和法律手续等,都是设计中必须考虑的要素。因此,建筑师转换观念,不将避难楼梯隐藏于后院,反而作为整个设施的重点积极地利用起来。

与自然的结合

从削减CO_2的排放、追求舒适自然共生型都市生活的理念出发,主楼梯的上部通过半透明玻璃可进行自然采光,而楼梯侧面与室外相连,形成一个自然通风的半室外空间。此外,屋顶安装庭园灌水专用的雨水收集设备及照明用的阳光板等。

与城市街区的结合

主楼梯与站前的交通环岛之间没有任何的障碍(如自动门等),与街道完全相通。沿着主楼梯走上去,你会发现这里有可以举办活动或是简单休憩的公共广场及随时感受四季变换的屋顶花园。一个平面展开的低层住宅却给人们带来了立体的视觉变化。

与铁路的结合

与之相邻的经堂站高架站台,比"经堂庭园"的2层还要高1.8米。为配合经堂庭园计划,将既有墙壁板改成玻璃幕墙。在现如今的日本都市空间大部分都只追求眼前利益,而在本项目中建筑师权衡站台、电车、主楼梯、

东侧立面
East facade

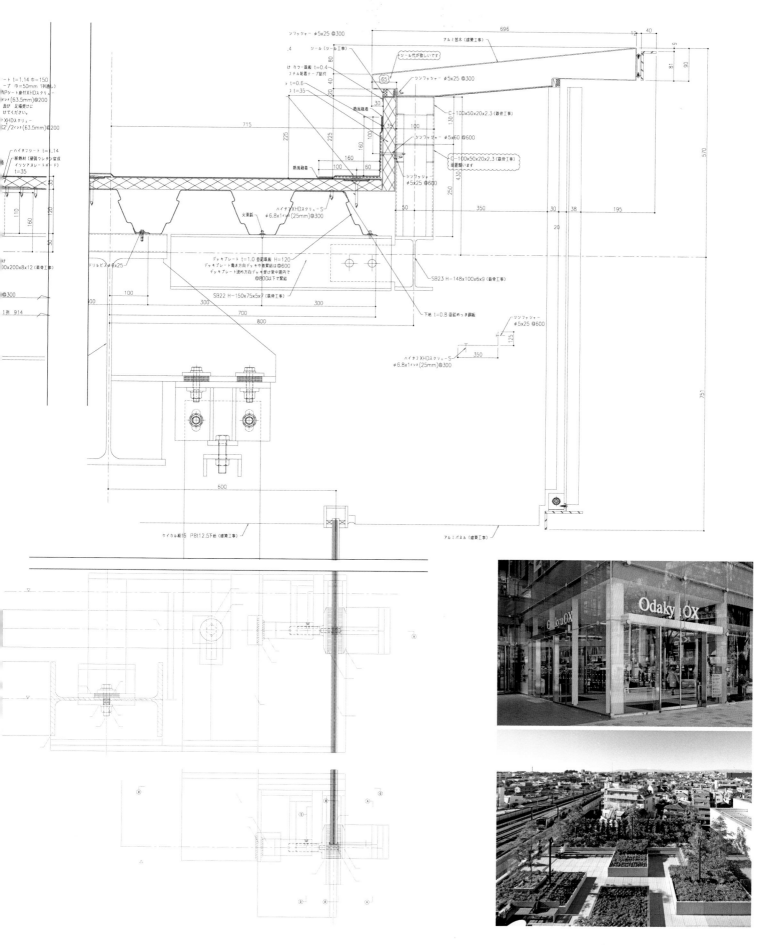

立面图
Elevation

東立面図

南立面図

沿街弧形北侧外观
Exterior of the north part

店铺之间的关系,不仅仅局限于对乘客的商业广告价值,而更重要的是展现出都市原有的魅力。2011年3月11日在日本发生的大地震引起的核电站泄漏事故,造成日本严重缺电,政府要求市民节约用电。而经堂庭园可自然采光、通风,因此得到了各界的高度评价。相信今后无论是日常生活还是无法预期的事态,经堂庭园将作为"都市核心"设施,永远发挥其作用。

KYODO CORTY is four-story commercial facility comprising 47 stores built in front of a station. Breaking away from conventional box-shaped commercial buildings, the goal was to combine nature, the city, the railroad, and commerce and create a facility that would form the core of the city. To this end, the architects did not hide the emergency stairs, which are required by the Building Standards Act, in the back, but rather actively used them as the main part of the architectural design of the facilities.

Nature

In light of reducing CO_2 emissions and pursuing an urban environment in harmony with nature, the rand stairs are set in a semi-outdoor space to let in

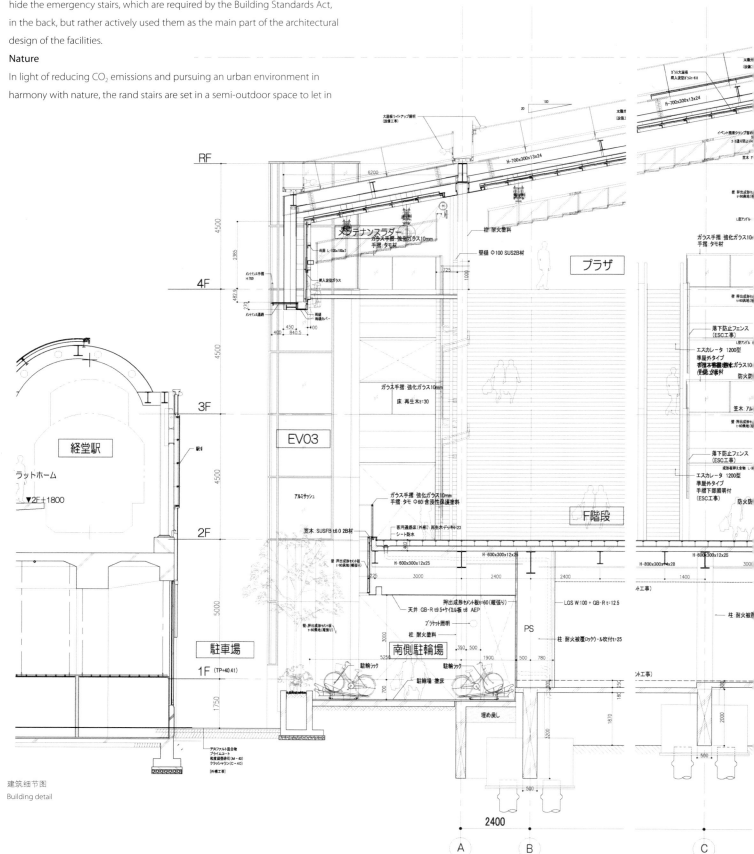

Building detail

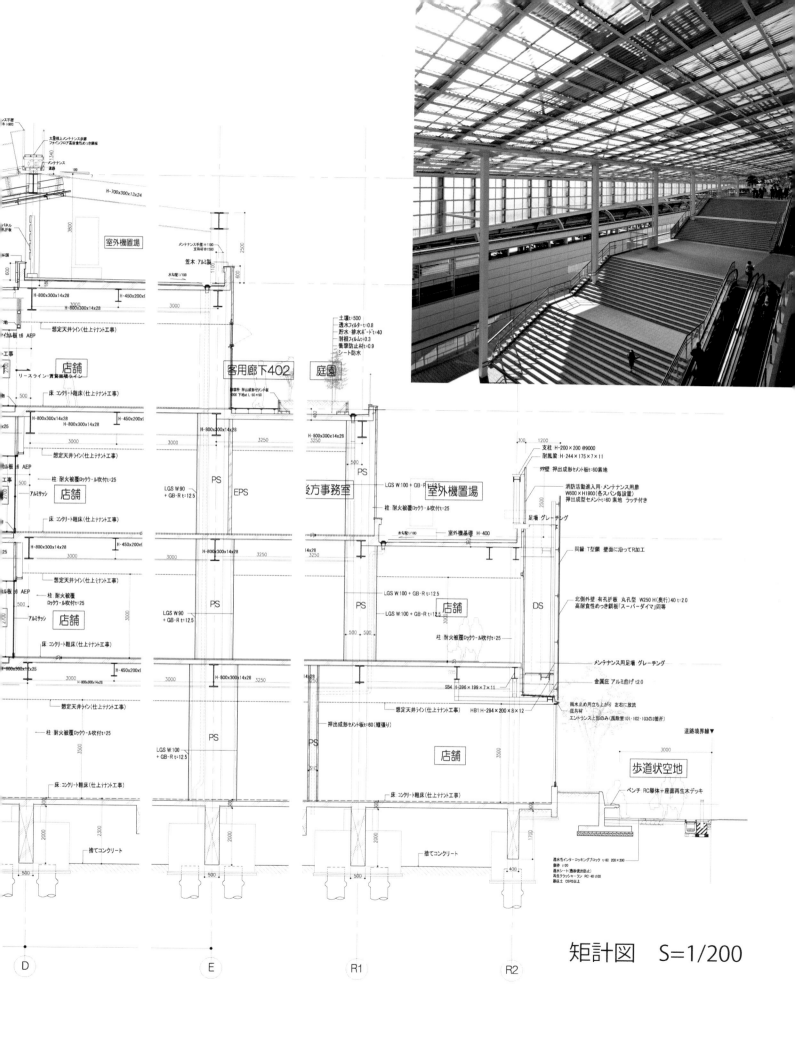

矩計図　S=1/200

立面图
Elevation

北立面図

西立面図

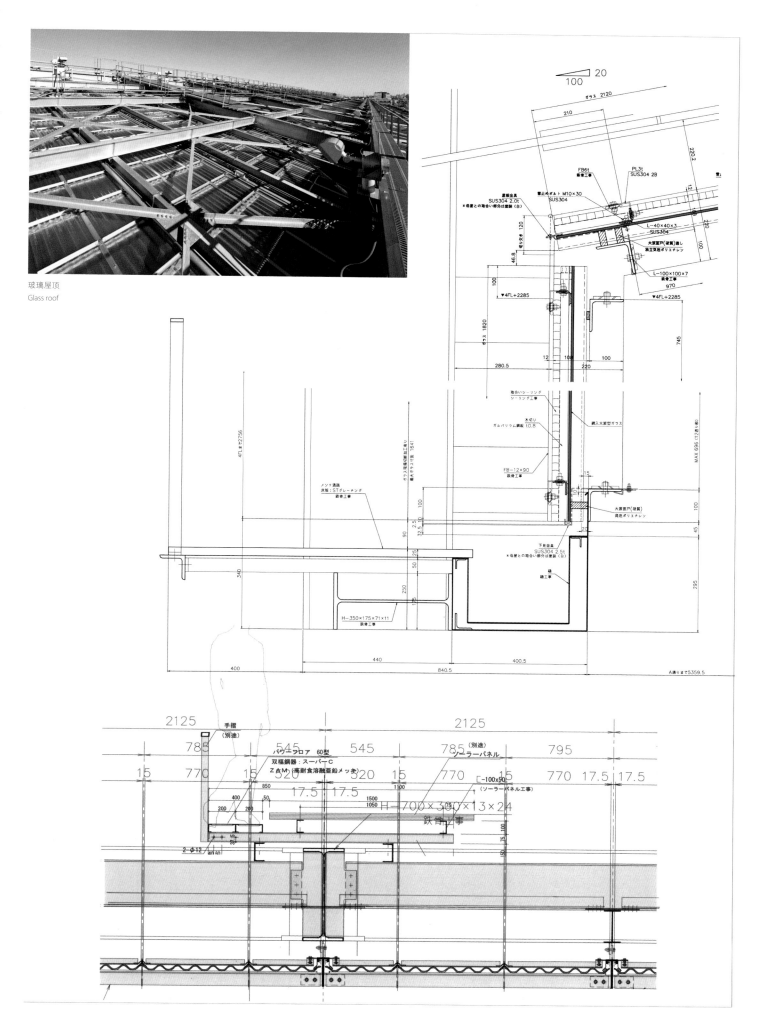

玻璃屋顶
Glass roof

natural light through a translucent glass panel installed in the upper part of the entrance and by opening the sides to the outside air for natural air circulation. They designed the roof to collect rainwater for garden irrigation and installed solar panels to light the common areas.

City

Because there are no doors or buildings between the grand stairs and the traffic circle of the station, the facility is open to the city and creates a seamless link from the station to the plaza and the roof garden on the top floor. Residents count on the facility as an evacuation center in times of disaster.

Railroad

In the midst of recent urban cities that overemphasize commercial efficiency resulting in closed spaces, the open relationships stretch over the platforms, trains, grand stairs, and stores, however small they may be in scale, to highlight the dynamism inherent in the city. It is expected that KYODO CORTY, a model facility that embodies the value of an energy-free new age, especially after the Great East Japan Earthquake, will become the core of the city.

Credits

Location: Tokyo, Japan
Site Area: 9,774.59 m²
Building Area: 5,433.33 m²
Total Floor Area: 15,650.90 m²
Designer: Shigeo YOKOTA
Photographer: Takahiro ARAI

一层平面图
First floor plan

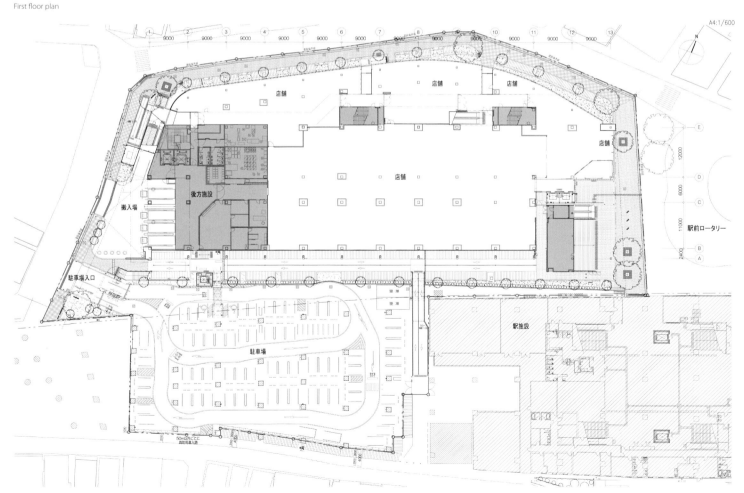

Center Stage | 中央舞台 |

SCI-Arc Graduation Pavilion
南加州建筑学院毕业作品展

| Oyler Wu Collaborative |

紧裹的幔布缠绕钢架整体，将其牢牢捻紧，使钢架宛如一处天然的布艺雕塑，幔布与锐利的钢框结构形成鲜明对比
The tightly stretched fabric twists around and through the structure, highlighting the natural curvature of the fabric in contrast to the sharpness of the structural frames

每年春天，南加州建筑学院都会挑选精干教职工团队和学生，为毕业作品展搭建场馆。2011年，Oyler Wu Collaborative 建筑事务所负责为该展搭建场馆，该场馆的不锈钢框架沿着12 192米的绳缆攀缘矗立，而绳索则将其紧紧系住，形成结构复杂的棚顶。今年，Oyler Wu Collaborative 建筑事务所再次被聘为毕业作品展的场馆设计，当然他们也面临着挑战：在维持去年设计的同时，他们需要融入对盛典的新构想。尤其需要指出，最大的挑战在于将已建展馆重新翻修。他们的应对策略分为两步：第一步着眼于毕业典礼的定位和进程，改变毕业学生的路线和典礼的视觉焦点。他们决定将观众席较去年来

个180度的大转向,去年这里面朝西侧的洛杉矶市中心,而今年面朝东侧400米长的学校大楼——被建筑学院称为"家"的地方;第二步,在观众席对面原先场馆的开阔区域,搭建了一处全新的舞台,被称为"中央舞台",是毕业典礼的举办地,还配备了发言代表的就座处。最主要的构件便是"中央舞台"的顶棚,它将与原先展馆雷同的建筑元素暴露出来,这次他们为宏伟的钢架配上了色彩缤纷的帷幔。

舞台如同不同构件互相交织的场所,而舞台的中心区则是它们融合的地方,环绕的座位形似露天看台,还有一处悬臂式的遮阳顶罩。舞台中心和原展馆中心完全对齐,而展馆整体却并不对称,因为露天看台和顶棚是偏离轴线的。

涂漆钢管搭建而成的结构几何造型源自舞台和露天看台,支撑着造型多变的舞台顶框。在浓密、弯曲的钢架内,天蓝色的帷幔从舞台台面延伸至悬臂顶端。紧裹的幔布缠绕钢架整体,将其牢牢捻紧,使钢架宛如一处天然的布艺雕塑,幔布与锐利的钢框结构形成鲜明对比。而今年的展馆最后一个翻新之处便是展馆本身。顶棚幔布由原先的银色变成了现在的炭灰色。这样的变化有效地转变了原先的色调——曾经的灰暗钢结构与亮色调的帷幔互相对立,如今银色的绳索在灰暗背景的映衬下熠熠生辉。

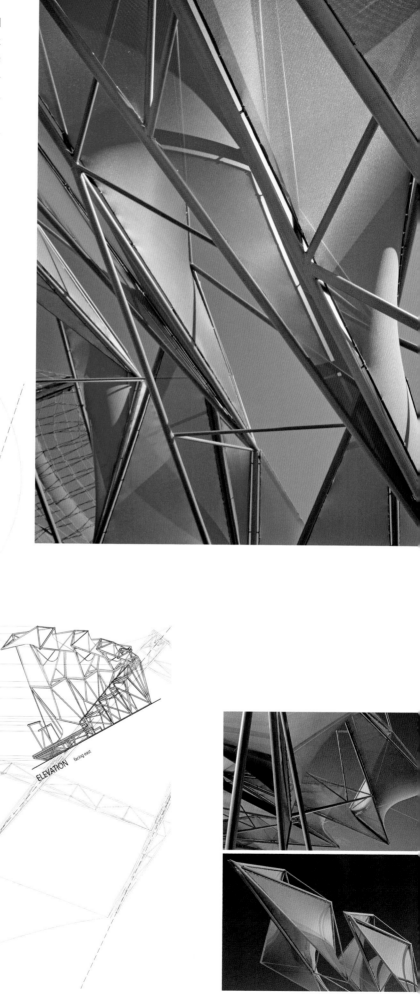

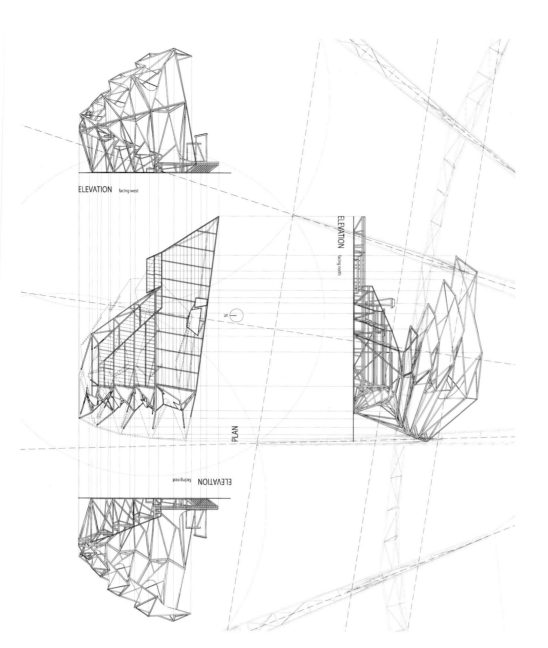

While the center of the actual stage is in alignment with the center of the existing pavilion, the overall structure is positioned asymmetrically, with the bleacher and canopy located off axis

Every spring SCI-Arc selects a faculty member, along with students, to design a pavilion for their graduation ceremony. In 2011, Oyler Wu Collaborative designed a pavilion constructed of a steel frame along with 40,000 linear feet of rope that was knitted to create an intricate canopy. This year, Oyler Wu Collaborative was again asked to design the architecture for the ceremony but with the challenge of rethinking the event of the ceremony while keeping the existing pavilion they had previously designed. Essentially, the challenge called for making the existing pavilion new again. Their approach to the problem was two-fold. The first strategy considered the orientation and flow of the ceremony by reworking the movement of the graduating students, as well as changing the visual focus of the ceremony. They decided to re-orient the audience 180 degrees from the previous year, turning them from facing west toward downtown Los Angeles to facing east toward the quarter-mile long school building that SCI-Arc calls home. Secondly, with the audience now facing outward toward the open end of the existing pavilion, a new stage, entitled Centerstage, was designed to formalize the celebration of the diploma ceremony as well as the seating of the guest speakers. The main feature, the canopy of Centerstage, plays off of the repetitious structural elements of the existing pavilion, this time combining an ambitious steel cantilever with a color twisting shade fabric.

The stage operates as a hybrid of different elements, incorporating into

Credits

Project Design and Fabrication Team: Dwayne Oyler, Jenny Wu, Huy Le, Sanjay Sukie, Mike Piscitello, Jie Yang, Clifford Ho, Tingting Lu, Mina Jun, Vincent Yeh, Justin Kim, Kubo Han, Sara Moomsaz, Amir Munoz, Tommy Shao, Mina Jun, Justin Kim, Sandra Reyes, Kathleen Mejia, Manuel Oh, Jim Li, David Ramirez
Structural Engineering: Nous Engineering, Matt Melnyk
Photographers: Scott Mayoral, Dwayne Oyler

it a large stage with a central podium, seating that is configured much like a bleacher, and a cantilevered shade canopy. While the center of the actual stage is in alignment with the center of the existing pavilion, the overall structure is positioned asymmetrically, with the bleacher and canopy located off axis. The structure is made of painted steel tubes that evolve geometrically from stage and bleacher supports into a twisting, repetitious frame that hovers over the stage. Within that dense and contorted frame, bright blue shade fabric winds its way from the stage surface up into the cantilever. Along the way, the tightly stretched fabric twists around and through the structure, highlighting the natural curvature of the fabric in contrast to the sharpness of the structural frames. The final element in the transformation of this year's ceremony is with the existing pavilion itself. The fabric of the existing shade canopy was changed from silver to charcoal grey. The change effectively inverts the colors of the existing pavilion—once reading as a network of dark lines set against a light field of fabric, the silver ropes are now highlighted against a dark backdrop.

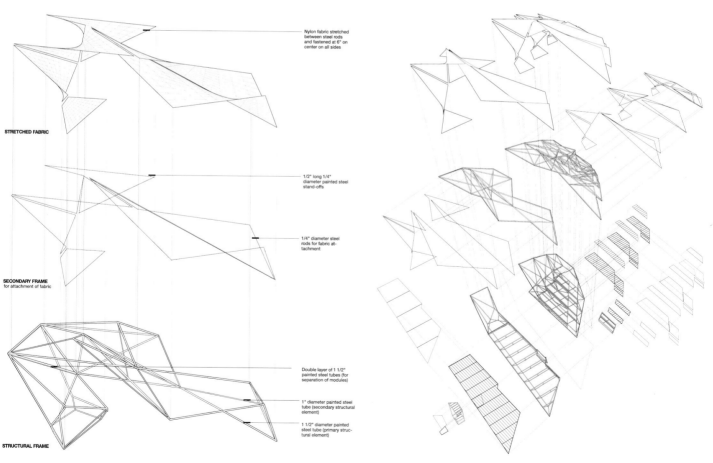

| Design Initiatives |

Jinzhou New Area Medical Center
锦州新区医疗中心

北侧，南侧立面图
North & south elevations

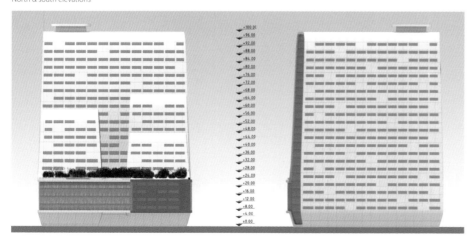

医疗中心的形态必须满足病患每天的需求。该方案旨在将锦州新区医疗中心建在已有医院的裙楼附近，从而达到缩短路程的效果，并使其与已建的医疗大楼融为一体。

建筑师已经设计了医疗中心的四座裙楼：分别为门诊楼、住院楼、医疗科研楼和教学/行政楼，四座大楼围绕一座步行广场而建。这座广场兼有大堂的作用，是多功能的室内外兼有的空间，并且介于走道和建筑之间，能够遮挡雨雪，又不至于被墙面全部封闭。四座建筑的小型入口厅堂向两层楼的广场打开，分散了人流，也让整个区域适宜行人散步休闲。

建筑师将小汽车通道和卡车通道分开，同时，还在围绕场地的两条主街上设置了两个下车/卸货点。

四座建筑之间互相分离，以便每个房间都能有自然光照和通风，这在医疗建筑当中是很需要的，而在18.5万平方米的建筑中做到这一点也不容易。大部分的病房都有着朝向西南部公园的良好景观视野。

除了通常的节能和节水设施、雨水和灰水的再利用以及材料的回收再生等，施工的密度也从标准的50%进一步减少了10%，从而减轻了建筑所占区域的影响。一半的地块专门用于公园和医疗中心后院的建设，后院原先属于公园的一部分，公园已经被拆除。还需要确保在中心周围两条大街上的公园入口通畅，因为目前还没有一条易行、便捷的公园通道。现有的建筑将会被拆除，用于建造公园，公园下还有两层停车场。可以确保中心区域的绿化覆盖率超过30%，而且除了公园占据的一半地块之外，建筑的第六层也成为绿色屋顶花园，这是与现有医院相连的"高层花园"。

建筑采用标准的10米×10米格栅（部分10米×11米）的钢结构建造，外部是金属板和幕墙，并穿越了高压电缆区。在施工过程中还广泛运用了预制构件。

东侧立面图
East elevation

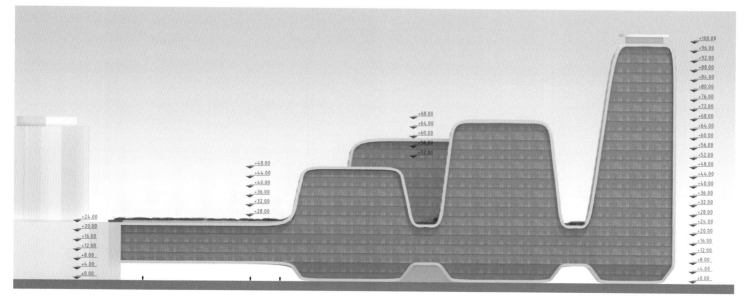

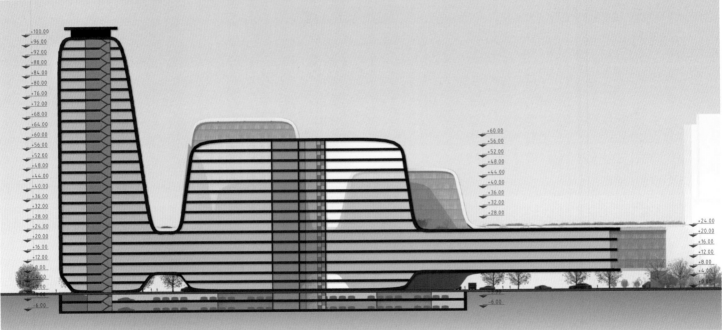

剖面图
Section

The everyday experience of the users is critically important for the hospital typology.

Our proposal for a Jinzhou New Area Medical Center is located as closest as possible to the existing wing of the hospital in order to shorten the routes and form one integrated complex with that existing wing.

We designed the four main wings of the Medical Center (Outpatients; Inpatients; Medical Tech; Teaching/Research/Administration) around a pedestrians only Lobby/Plaza—a multifunctional, transitional, between indoor and outdoor, between the sidewalk and the building spine which is covered in protection from rain and snowfall but is not enclosed with walls. The small entry lobbies of the four wings are open to that two story Hub/Plaza which distributes the crowd and makes the whole area more pedestrian friendly.

We split cars access from trucks access and provided two drop off areas at the two main streets fencing the site.

We shifted and offset the four different wings so every room that needs it has an access to natural light and ventilation—something so rear in medical buildings and hard to get organized with an area of 185,000 m². We oriented most of the wards towards the best view—the park (south-west).

In addition to the usual energy and water conservation, rain water and grey water use, incorporation in construction process of recycled and reused materials, etc. we provided 10 % less than required 50% Construction Density in order to minimize even further the impact of our building's footprint. This way over a half of the lot is dedicated to the Park and The Medical Center backyard becomes a part of the existing and now neglected Park. We ensure park entries from the 2 big streets surrounding the Center because now there is no easy and direct access to the Park. The existing buildings on the site are to be demolished

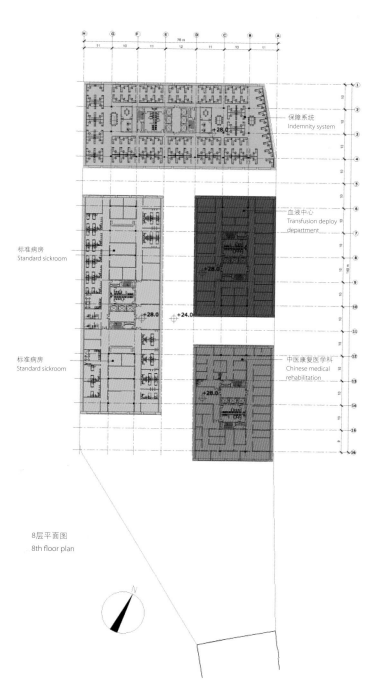

8层平面图
8th floor plan

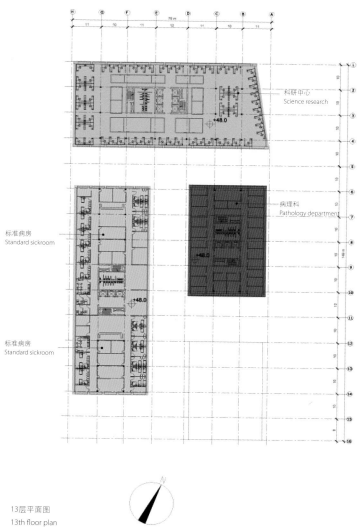

13层平面图
13th floor plan

Credits

Location: Dalian, China
Area: 184,828 m²
Date: August, 2012
Type: Healthcare
Status: Competition
Client: First Affiliated Hospital of Dalian Medical University
Project Team: Vlado Valkof – project architect;
Ana Valkof, Stefan Petkov, Malgorzata Blasik, Minko Marinov – architects;
Nick Tonchev – civil and structural

and clear space for the park on top of the 2 underground levels garage. We significantly exceed the required 30% Greenfield Rate and in addition of the half lot park we provide a green roof garden on top of the 6th floor—an elevated park which connects with the existing flat roof of the existing hospital.

It is going to be a steel structure on standard 10x10 m grid (10x11 m in some cases), metal panels and curtain wall for the envelope, and space frame for the bridge over the high voltage cable area. Prefab elements will be used widely in the construction process.

Natura's Showroom Santo André
Natura圣安德烈分店

FGMF

入口
Main entrance view

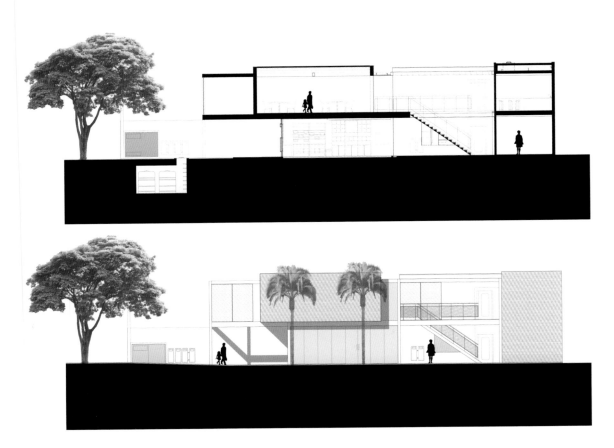

花园里的楼梯
View of the stairs from the garden

室内
Inside view

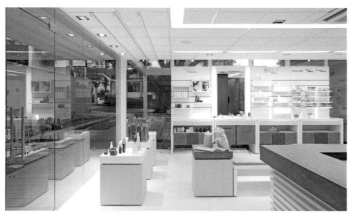

室内
Inside view

在着手Natura圣安德烈项目之前，建筑师们先回顾了一下为Natura在圣保罗设计的5个分店。这些地方专为Natura这一巴西最大的化妆品公司设计，该公司拥有大约100万直销人员。

设计关注的重点是空间品质和流动性，功能的划分和安排不是该项目的全部。除了在视觉交流和展示区植入一丝不苟的品牌策略外，也着重体现了其空间品质和流动性。

考虑到项目施工进度和进程控制，这个工程在异常干燥的环境下进行（这在巴西十分罕见），一切就绪后，降低了出错和重做的几率。减少浪费，使用再生或可再生材料方面也做得很好。

只有在地形和地基处理上使用了传统方法。每个原件和元素都依次运入基地并组装：金属结构、石膏墙板、胶合板、框架和辅助的装修材料。这也是该项目的亮点所在，被FundaçãoVanzolini/POLI-USP授予AQUA（高品质环境）环境认证奖。AQUA是继法国的HQE之后第一个完全适应巴西实际情况的认证体系，也是享有国际最高认证资格的"可持续建筑联盟"的成员。"Natura圣安德烈分店"是巴西第一个获得AQUA认证的商业项目。

还有其他一些值得一提的环境策略：施工尽量避免对邻居产生消极影响，并为城市提供了一个慷慨的花园；材料的选择易于维护，建筑充满了有趣的方案。这些大多数是建筑师在项目进行中想到的，如自然光、阳光控制和地域的高渗透性。此外，开放式的设计使房间在没有空调时也可实现通风。建筑师还选择了易于维护的材料，如着色铝外墙只需要清水清洗，室外地板使用了再生材料。这还牵扯到水和能源的管理；湿热、声环境、视环境和嗅觉的舒适性；花园管理和其他。

根据要求，该建筑需要适应巴西不同的地区，因此在建筑结构中考虑了

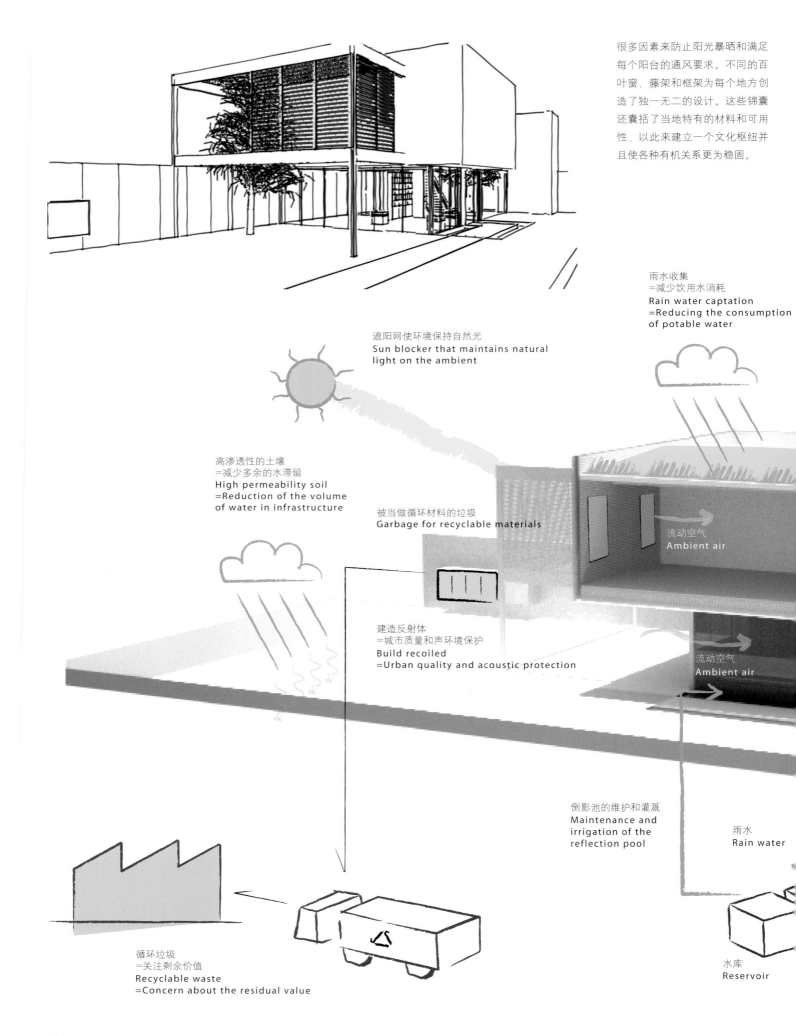

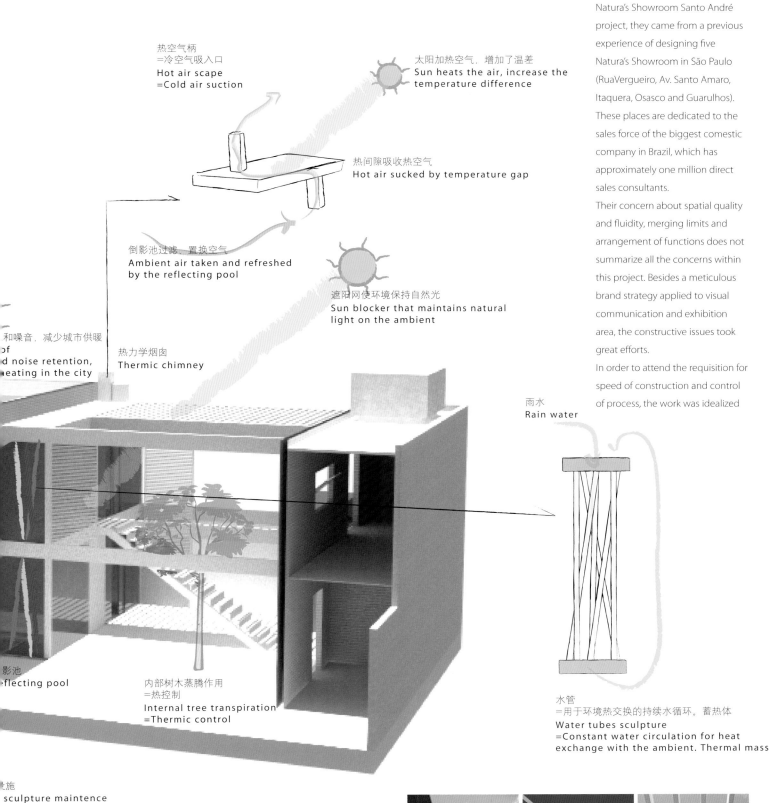

热空气柄 =冷空气吸入口
Hot air scape =Cold air suction

太阳加热空气，增加了温差
Sun heats the air, increase the temperature difference

热间隙吸收热空气
Hot air sucked by temperature gap

倒影池过滤，置换空气
Ambient air taken and refreshed by the reflecting pool

遮阳网使环境保持自然光
Sun blocker that maintains natural light on the ambient

和噪音，减少城市供暖
of noise retention, eating in the city

热力学烟囱
Thermic chimney

雨水
Rain water

影池
flecting pool

内部树木蒸腾作用 =热控制
Internal tree transpiration =Thermic control

水管 =用于环境热交换的持续水循环。蓄热体
Water tubes sculpture =Constant water circulation for heat exchange with the ambient. Thermal mass

施
sculpture maintence

When the architects first began the Natura's Showroom Santo André project, they came from a previous experience of designing five Natura's Showroom in São Paulo (RuaVergueiro, Av. Santo Amaro, Itaquera, Osasco and Guarulhos). These places are dedicated to the sales force of the biggest comestic company in Brazil, which has approximately one million direct sales consultants.

Their concern about spatial quality and fluidity, merging limits and arrangement of functions does not summarize all the concerns within this project. Besides a meticulous brand strategy applied to visual communication and exhibition area, the constructive issues took great efforts.

In order to attend the requisition for speed of construction and control of process, the work was idealized

国际新建筑 **133**

View of the garden

to be obsessively dry (very rare in Brazil's construction scenario), all set, reducing the chance of errors and need for remakes. The concern about reducing waste, using recycled or recyclable material is also plainly fulfilled by this option. Only the preliminary work on terrain and foundations were executed by traditional methods. On sequential steps, every component and element came to the building site divided and then were assembled: metallic structure, plaster wallboards, cementitious boards, frames and complementary finishings. This matter was one of the highlights on the project for environmental certification attainment AQUA (High Environmental Quality), granted by FundaçãoVanzolini/POLI-USP. Created after the French seal HQE, AQUA is the first seal fully adapted to Brazilian reality, and member of the Sustainable Building Alliance, which congregates international greatest certifications. The Natura House Santo André project is the first commercial project certified by AQUA in Brazil.

There are other environmental strategies that must be mentioned: From the caution on the implementation of the House, avoiding negative impact on neighbors and offering a generous garden to the city, to the choice for certified material with easy maintenance, the building is filled with interesting solutions. Most of them were thought while still on project, present since the first strokes, such as the concern for natural light, sunlight control and the high permeability of the terrain. Or else, the opening system which permits crossed ventilation on days when air conditioning is not essential. Or, still an example, the option for material that reduce future maintenance, such as painted aluminium external coating, which demands only washing, and external floor made with recycled material. The matter is extended to the management of water and energy; hygro-thermal, acoustic, visual and olfactive comfort; garden management and else.

Accordingly to the requirement that the house is adjustable to different regions of Brazil, kits of elements were conceived, which may be attached to the House's structure as needed due to insolation and ventilation variations of each terrain. Different kinds of brise-soleil, arbors and frames that may compose a singular performance and aspect design for each location. These kits also take in account typical material and the availability of local industry, in a way to also build a cultural connection and to stablish more organic relation with each scenario.

结构细部
Structural detail

室内
Inside View

上层平面图
Upper Floor Plan

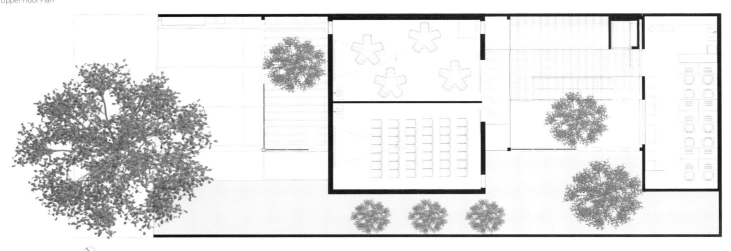

135

外部空间夜景
Night view of the exterior area

产品展示区
View of the product's display

Credits

Authors: Marcelo Bicudo, Fernando Forte, Lourenço Gimenes and Rodrigo Marcondes Ferraz
Coordinators: Tatiana Machado and Ana Paula Barbosa
Architects: Naya Adam, Juliana Nohara
Interns: Flavio Faggion
Structure: OppéEngineering
Landscaping: Studio Ilex
Lighting: MarcosCastileLighting
Electric: PexElectricalProjects
Construction: BR CONSTRUCTIONS
Joinery: Mão Colorida
AQUA Consulting: ProActive
Air Conditioning: LFB ThermalEngineering
Photographers: Fran Parente and Demian Golovaty

室内
Inside View

主平面图
Main floor Plan

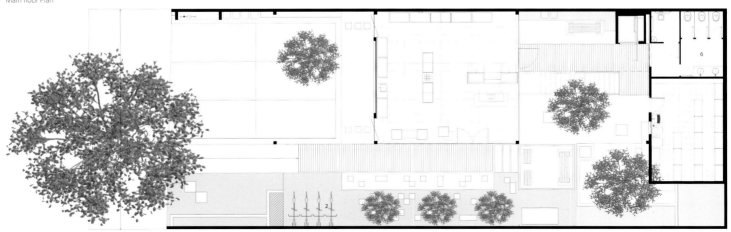

西视角
Exterior perspective—west

| AB Architecken |

Cheung Residence | 张氏别墅 |

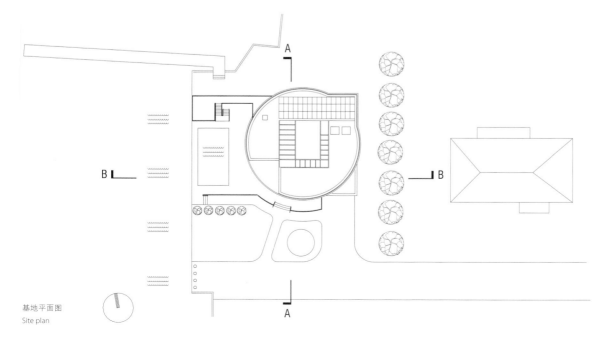

基地平面图
Site plan

该住宅的设计灵感来源于一枚中国的古代铜钱。外表面是圆形格子变色玻璃气窗,提供了遮阳的同时也反射了周围的自然环境,也暴露了内部沿中心庭院向上攀升的混凝土结构。天井和倒影池将光线带入内部,并且使得整个住宅对流通风。西南面的另一个垂直镂空既可作为私家花园,又是通向内部的缓冲区。

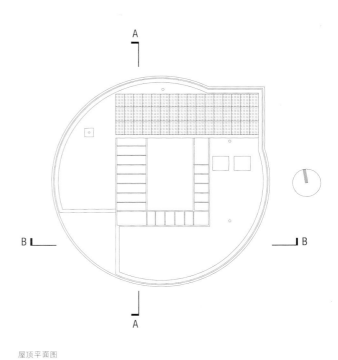

屋顶平面图
Roof Plan

剖面A
Section A

剖面B
Section B

西立面
West elevation

北立面
North elevation

入口东南视角
Entry perspective—southeast

西南面的客厅室内
Interior perspective—living room—southwest

Inspiration for the design of this house was drawn from the form of an ancient Chinese coin. A circular lattice of photochromic glass louvers forms the exterior shell, providing solar shading and reflecting the natural context. It veils the concrete structure within, which ascends in section around the central courtyard. The courtyard and reflecting pool bring light into the interior and allow cross-ventilation throughout the house. A second vertical void, located in the southwest sector, doubles as private garden and environmental buffer to the main living area.

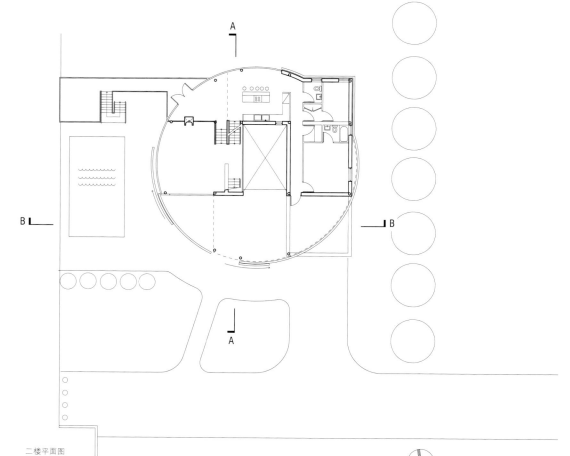

二楼平面图
Second floor plan

西南视角
Exterior perspective—southwest

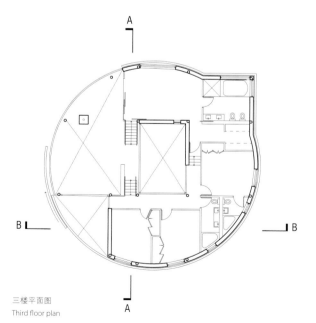

三楼平面图
Third floor plan

Credits

Location: New York, NY, USA
Area: 334m²
Type: Single-family Residence

西北视角
Exterior perspective—northwest

从停车场看总部大楼，大楼周围种着80种树
View from visitor carparking. There are over 80 different types of trees planted on the site

Jansen Campus: Building A Vision

Jansen总部大楼：树立愿景

| Davide Macullo Architects |

初始概念草图
Original concept sketch

Jansen总部大楼——架在地区基因及其未来之间的桥梁

新建的Jansen总部大楼选址位于一片工业区的北侧,与村庄的一处扩建的小型居民区交界。由于地理位置特别,新建大楼连接了两处不同的城市聚居区——大楼既是工业区的大门,也通过减少规模与村庄相协调。建筑师通过将大楼建筑一分为四,从而达到缩小建筑规模的目标。

在Jansen总部大楼的设计中,建筑师致力于创新型材料和高科技方案的研究,部分材料和科技还是首次用于施工中。比如,由Jansen独创的半结构化建筑外立面就是一套全新的体系,在无需外部支撑体系的前提下,保证了大楼内光照、玻璃以及透明构件的延续性。

为了建造大楼平缓的屋顶,一套在混凝土铸模中加入纤维的系统应运而生。设计方通过该方法保证了灌注的水泥能够加固大楼的金属结构。

棱角分明的屋檐为室内空间阻挡了夏日的骄阳
Sharp overhangs protect the interior spaces from high summer sun

屋顶如周围的山脉一样起伏
The sloping roofs rise and fall like the surrounding mountains

Jansen还参与了创新性的辐射系统的设计，系统基于热辐射原理而建，已经被安装于大楼之内；而大楼地板、天花板的混凝土结构内部直接安装了制冷、制热管道，确保了楼内高质量的环境。

大楼外立面覆有暗色调的孔状钛锌板。建材经过独特的装饰着色，烘托了周边木结构建筑的紧致。作为首度使用的外部镀层，钛锌板表面的光泽在一天的不同时间时明时暗。组合式的设计和坚韧的钛锌板在大楼整体设计中举足轻重，使靠近的游客感到妙趣横生。

Jansen总部大楼里外外都用实用的资源打造而成，这些资源在不远的周边都可购得。这样的建筑也体现了该地区的企业优势，对环保建筑原则的提倡以及对节能增效的关注。

内部构造

为了使员工在大楼内的日常工作顺利进行，公用空间被设计在主电梯和楼道的附近，而更私人的工作区则沿着大楼的通道而建。大楼的结构功能被三角形的外墙限定，因此使内部空间的安排更有灵活性。现如今大楼空间被规划成三维网格形态，迎合了公司本身的功能架构。底层的接待区是大楼的公共区域，区内有会议室、商务午餐会所和餐厅。接待区旁边则是以"任务控制"著称的办公室，这里是公司的心脏，氛围几乎和证券交易所相同。所有关于公司运作的信息都在那里经过汇总并加工处理后上报。位于一楼的是名为"Kreativbereich"的空间——全开放的办公空间和信息会议室备受雇员们的亲睐，正门处还有教学室和其他会议室。通信区的办公室位于二楼，三楼则是靠近全景观景台的会议室。

沿着盘旋的楼梯分布着个人办公间和私人工作

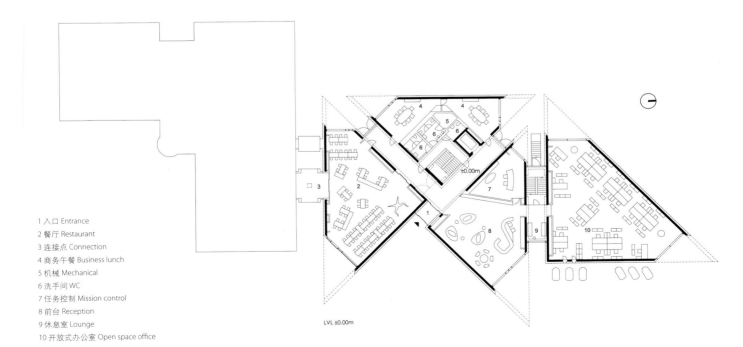

1 入口 Entrance
2 餐厅 Restaurant
3 连接点 Connection
4 商务午餐 Business lunch
5 机械 Mechanical
6 洗手间 WC
7 任务控制 Mission control
8 前台 Reception
9 休息室 Lounge
10 开放式办公室 Open space office

建筑的规模、材料以及倾斜的屋顶都是对当地住宅环境的致敬以及重新诠释
The scale, material and the sloping roofs respect and reinterpret the local residential context

从老工厂的楼顶上望去。公司的工业遗产在新的总部大楼中体现了出来
View from the roof of the existing factory. The industrial heritage of the company is reflected in the materials and forms of the new HQ

东入口立面,钻石状的窗户覆上了莱茵辛克网
East entrance facade. The diamond windows are covered with Rhinezink mesh

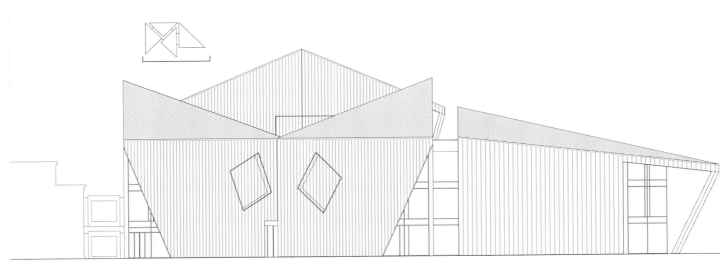

东侧视图
East view

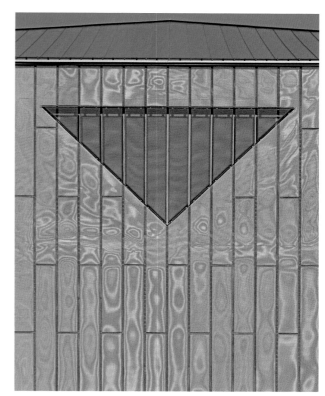

闪亮的莱茵辛克立面
The shimmering Rhinezink facade

区，这里看上去更加幽静，越往上走，这样的区域也就越多。位于最北侧的三角体建筑内的是公司的生产线，占据了大楼的两个楼层，而生产部门管理人员的办公室位于楼上。南侧的三角体建筑是公司的质检区，相关人员在二楼的办公室内管理区域运作。地下室内1 000平方米的区域坐落着文献室、机械操作间和技术设备区。

虽然这里地形复杂，但建筑的内部环境将这一负面效应降至最低。建筑构件因此十分惹眼，遵循了项目和选址理性、经济的开展思路，也使空间更具技术和工业气息。

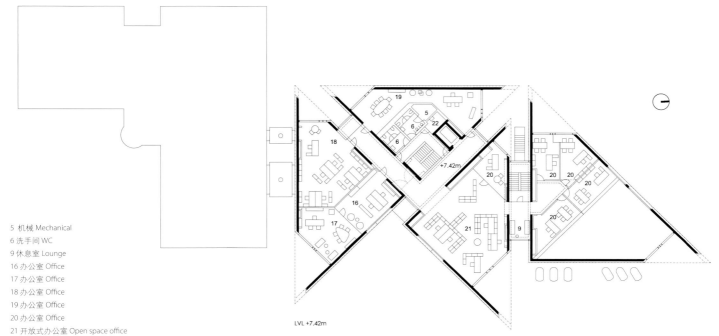

5 机械 Mechanical
6 洗手间 WC
9 休息室 Lounge
16 办公室 Office
17 办公室 Office
18 办公室 Office
19 办公室 Office
20 办公室 Office
21 开放式办公室 Open space office
22 维护 Maintenance

LVL +7.42m

夜景。水池可用于雨水收集
Night view. The pools serve for rainwater collection

从老工厂看新的办公总部
View of the new HQ from the factory

倾斜的屋顶塑造了有趣的光影效果，视线越过屋顶可一瞥天空美景
The sloping roofs work together to make an interesting game of shadows and offer glimpses of the landscape

从室内空间还可以欣赏到莱茵河谷的秀美景色
The interior spaces are given sliced views of the Rheintal landscape

楼梯细部
Stair detail

Jansen Campus—A Bridge Between the DNA of a Place and Its Future

The site for the construction of the new Jansen Campus lies at the north end of the industrial complex and is bordered by the small scaled residential expansion of the village. This particular site allows the new building to insert itself as the link between two different urban scales, at once acting as the face of the industrial area while also reducing to the scale of the village. This reduction in scale has been achieved by fragmenting the mass of the building into four.

The new Jansen Campus is also characterized by research, carried out during the design, on innovative materials and technological solutions, some used for the first time in construction. For example the semi-structural facade, produced by Jansen, is a new system produced in such a way as to guarantee a continuity of the reflective, glazed and transparent elements of the building, without the need for external support mechanisms.

In order to build the sloping roofs of the building, a system of adding fibers to the concrete casting was developed. By doing this, this guaranteed that the poured cement would adhere to the metal reinforcements. An innovative radiant system

(TABS), partly produced by Jansen, based on thermal mass principles, has also been integrated into the structure; heating and cooling circuits have been installed directly into the concrete structure forming the floors and ceilings, ensuring the quality conditioning of all spaces.

The facade is clad in a dark pre-patinated perforated Rheinzink mesh. This particular finish gives the material a coloring that evokes the density of the tones of the wooden buildings of the surrounding area. Used for the first time as an external cladding, this shimmers with reflections and shadows, changing throughout the day. The modular design and the tight stretched mesh play a role in the scale of the building and make it interesting and pleasurable for approaching visitors.

The Jansen Campus, both internally and externally was almost entirely built using resources available within a few kilometers of the site. This fact highlights the entrepreneurial strength of the region, the commitment to sustainability principles and the focus of efforts towards effective energy savings.

Internal Fuctions

In order to allow for the fluid flow of daily working life, spaces intended for collective use have been placed adjacent to the main lifts and stair while the more intimate working spaces lie further along from this circulation. The structural functions of the building are assumed by the perimeter walls of the triangles, thus allowing for a free plan internally with a high degree of flexibility and possibility for future division. Currently the spaces are organized about a three-dimensional grid that corresponds to the company's functional structure.

会议室可谓艺术收藏室，汇聚了当代年轻艺术家的各色作品
The boardroom. The project also saw the beginning of an art collection. All the pieces in the building are by young contemporary artists

7米长的桌子是由建筑师亲手设计的
The 7m corian long table was designed by the architects

户外的公园与雨水收集池
Exterior view, the park, the rainwater ponds

办公区是特别定制的。墙上的"iSacpes"字样是由建筑师设计的
Work spaces were custom designed. The prints "iSacpes" on the wall are by the architect

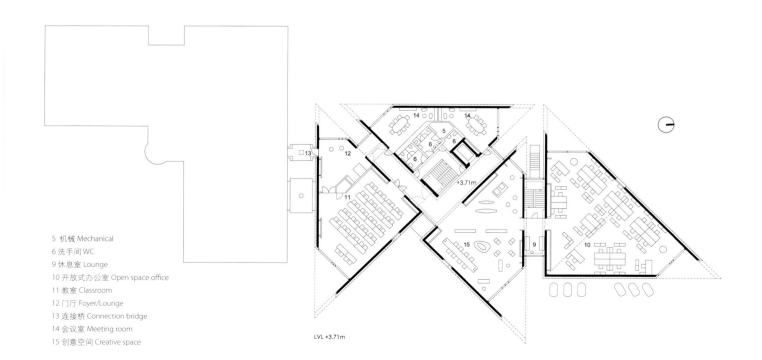

5 机械 Mechanical
6 洗手间 WC
9 休息室 Lounge
10 开放式办公室 Open space office
11 教室 Classroom
12 门厅 Foyer/Lounge
13 连接桥 Connection bridge
14 会议室 Meeting room
15 创意空间 Creative space

LVL +3.71m

The public functions are distributed from a reception zone on the ground floor. Rooms for meetings, business lunches and a restaurant all lead off this area. Also on the ground floor, beside the reception is an office known as "Mission Control" representing the operational heart of the company and acts almost like the stock market floor, where all information regarding the operations of the company is processed here in real time. On the first floor there is a space named "Kreativbereich", a workplace and informal meeting space open to all, much appreciated by the employees, a teaching room with foyer and other meeting rooms. An open plan office for the communications section is located on the second floor and on the third is the boardroom with a panoramic terrace. Individual offices and more intimate working spaces requiring more privacy are distributed along a spiral, with their area increasing as the spiral rises. The northern-most triangular block houses the operations wing of the company across two floors and on the upper floors are the offices of the directors responsible for this sector. The south triangle houses quality control and the executives responsible on the second floor. In the basement there are ca. 1,000 m² reserved for archives, mechanical rooms and technological systems.

Despite its apparent sophistication, the atmosphere of the internal landscape reflects the principle of reducing details to a minimum. The constructive elements are therefore always explicit and follow the rationale and economy of the site and the project, giving the space a technical, industrial atmosphere.

楼梯细部
Stair detail

悬空式屋顶上安装了照明、防火、振动检测、电力和电缆系统、声音和空气调节器。与传统的无梁式天花板不同的是，这些设施更加安全，而且装饰效果更佳

The acoustic suspended ceilings contain all the lighting, fireprotection, movement detection, electrical and cabling systems, audio and air conditioning. Unlike a traditional flat ceiling, these are much lighter, decorative elements that escort you around the building

夜晚,灯光透过网状的窗格洒向室外
Night view. Light glows out from behind the mesh window screens

从平面到立面各个角度都能看到三角造型
The trianglar geometry can be seen in both plan and elevation

Credits

Location: Oberriet, SG, Switzerland
Project Start Date: July 2008
Construction Start Date: May 2010
Completion Date: May 2012
Site Area: 3,705 m²
Building Area: 1,100 m²
Total Floor Area: 3,300 m²
Basement Floor Area: 900 m²
Above Ground Floor Area: 2,400 m²
Volume: 15,800 m³
Stories: 1 Level Basement, 4 Levels above Ground
Function: Office Building
Certification: Minergie Label
Client: Jansen AG
Photographer: Pino Musi, Enrico Cano

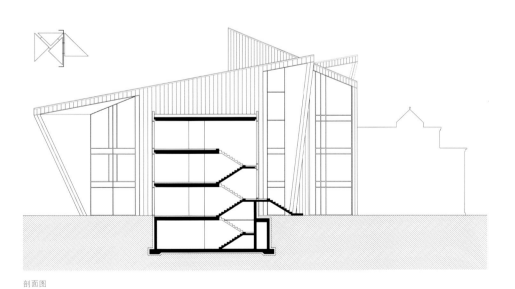

剖面图
Section

陈列馆与主展厅之间的波浪饰面板如瀑布般倾泻而下
Cascading timber veneer panels separate the gallery and main exhibition hall

奥克兰交通科技博物馆位于奥克兰Meola路，新的飞机展示厅是对原有建筑的扩建，是整个博物馆重要的展览空间。占地3 300平方米的新建空间将使博物馆能够容纳所有的飞机展示品，这其中包括一些非常珍贵的飞机。展厅全部采用木材建造。鉴于机库的面积大小，大厅完全由预制的层积木打造而成，同时翼展采用层积木门式钢架建造。展厅内宽42米，是新西兰所有层积材木结构中最宽的建筑。

飞机展示厅作为展示飞机的窗口，内部是一个大展厅，犹如一个"黑匣子"，而外部则直观地体现了大楼面朝大街的形态和模样。大楼北侧玻璃外立面自然通风，像是在挑檐下方浮动一般，展现出建筑和后方走廊的格调美。两根立柱矗立于入口两侧，营造出光线明亮的二层空间，同时这两根立柱也是12米高的热烟囱，满足了自然通风的需求。而在展厅南部基本展览区的夹层、讲堂和服务空间处也采用了双柱式入口的设计。

通过与所拥有展品的技术互相辉映，使展厅建筑的构造得以表达。建筑内衬由木制胶合板构成，基于环保理念设计，迎合了展厅整体的木结构，为大楼注入了柔情和高雅，并使待展的飞机和展示背景之间形成鲜明对比。从展示厅设计一开始，可持续性的设计理念便成为整个设计方案的重要部分。而从展示厅落成开始，它便已经获得数项可持续设计大奖。展示厅坐落地原是一处垃圾填埋场，有效地利用了一块回收用地。从大厅高处到底部，到处可见利用传动装置操控的窗户和通风口，用于自然采风，这意味着空调在这里无用武之地。而设计过程中，生命周期研究理论被运用于建筑材料的选择过程中，这包括了对碳排放平衡和对各种建筑结构内能的实证研究。层积木材料由此被运用于展厅设计中。而雨水灌溉系统也被引入，协助洗手间的日常运作。而为了营造最佳的场地环境，设计方还制定了瓦斯管理战略。

北立面
Main northern facade

MOTAT Aviation Display Hall
奥克兰交通科技博物馆飞机展示厅
| Studio Pacific Architecture |

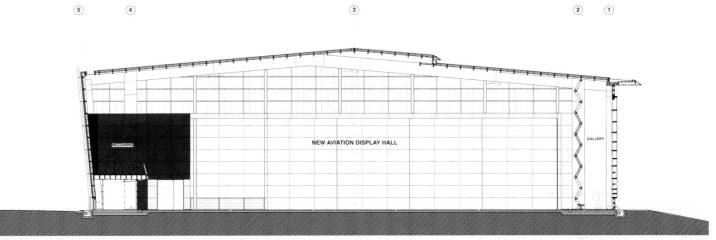

北立面玻璃幕墙,复合有机玻璃细部
Detail of the combined plexiglass and glass northern facade

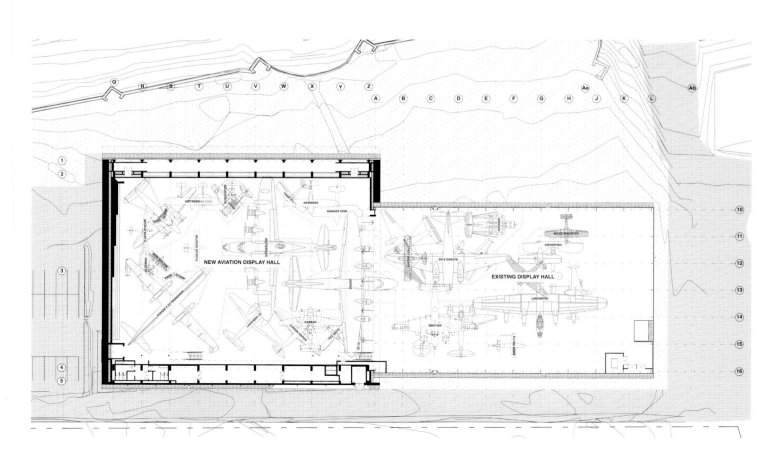

The Aviation Display Hall is a major new museum facility for the Museum of Transport and Technology (MOTAT)'s collection of historic aircraft. It was designed to extend the existing aviation building on the museum's Meola Road site in Auckland. At 3,300 m², the extension will enable MOTAT to house all of their aviation exhibits, some of which are the only remaining examples in the world, in a protected environment for the first time.

The building's design has been resolved as a timber interpretation of the hangar form. Unusually for its size, the structure is almost entirely fabricated from Laminated Veneer Lumber (LVL), and uses the unique capacities of LVL portal frames to encompass the large wingspan of the aircraft. At 42 meters internal width, the building has the largest clear span of any LVL timber structure in New Zealand.

Designed to showcase the historic aircraft, the Aviation Display Hall houses an inwardly-focused "black box" exhibition space, while the exterior provides visual expression of the building's form and contents from the street frontage. The naturally ventilated northern façade glazing, floating beneath the projecting roof, reveals the rhythm of the structure and the gallery wall beyond. The double legged portals are exploited to create a second light-filled gallery space on the north of building that also acts as a 12 m high heat chimney, supporting the natural ventilation strategy. On the south of the building these

主展厅内部
Interior of the main exhibition hall

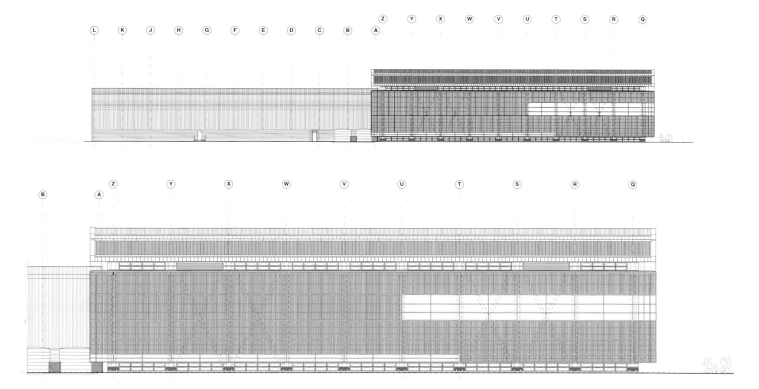

12米高的陈列馆同时也扮演着太阳能烟囱的角色
The 12 meter high gallery space acts as a solar chimney

double legs are utilized to locate the mezzanine, classroom and support spaces within the primary exhibition space.

The tectonics of the building's structure is expressed in a manner that reflects the technology of the exhibits it houses. Timber veneer linings, selected on a sustainable basis, echo the materiality of the timber structure, bringing warmth and grace to the building and providing a contrasting textural background to the aircraft.

From the outset, sustainable design measures have been an integral part of the overall design approach, and since its completion, the Aviation Display Hall has won a number of Sustainability awards. Situated on a former landfill, the building effectively makes use of recycled land. High and low level banks of actuator controlled windows and louvers throughout the building provide natural ventilation, meaning that no mechanical air conditioning is required. Life cycle analysis, including empirical research into relative carbon balance and embodied energy of various structural options, was commissioned prior to structural material selection. This supported the use of LVL in the building. A rainwater harvesting system is employed to service the toilet facilities, and gas management strategies were also implemented in response to prevailing site conditions.

Credits

Location: Auckland, New Zealand
Completion Date: June 2011
Area: 3,300 m²
Materials: Laminated Veneer Lumber (LVL) Structure, Polycarbonate Glazing, Décortech internal Timber Veneer Paneling
Project Team: Evzen Novak, Marcellus Lilley, Grant Perry, Anna Windsor, Brendan Himona, Mark Hadfield
Client: Museum of Transport and Technology (MOTAT)
Photographers: Patrick Reynolds, Matt Wilmot

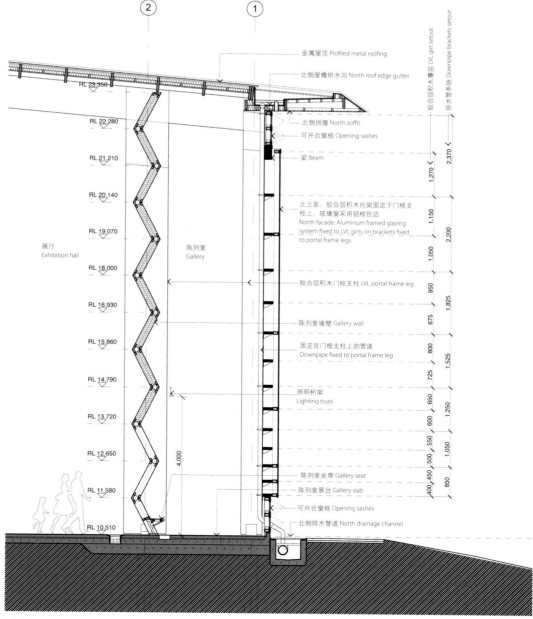

North facade

作为住宅的最远端，平台在人行道上方不到两米，最外围是一个花坛，小型灌木也可以使建筑看上去更小
At its farthest extremity from the house the deck sits less than two meters above the sidewalk. The facade is further broken down by the planter box, adorned with small shrubs which help to reduce the scale of the building

Weiss Residence ｜韦斯住宅｜

| Robert Harvey Oshatz Architect |

韦斯住宅坐落在俄勒冈州波特兰郊区西瀑布山的一个斜坡上，提供了一个壮观的街景。它严格的几何形和不寻常的细节设计突出了这是一个形状造就的有趣组合。尽管这不是罗伯特·奥赫兹最精美的作品，但可能是他最令人钦佩的。韦斯住宅是奥赫兹为他的女儿女婿设计的，属于这对夫妇和他们孩子的第一个经济房。受到经费的限制，能在这个地势险峻的地方建这样一个独特的住宅实在令人啧啧称奇。

这个179平方米的建筑有三层高，使住宅给人一种垂直的印象，东西立面类似塔楼造型，但从街道上看去并不会显得很突兀。住宅入口在中间层，在一个向下的混凝土短车道后面，屋檐向下倾斜形成钻石状，以至于上层楼几乎被隐藏起来。住宅通过滑动门向一个大型矩形平台开放。平台扮演了传统门廊的角色，帮助居住者融入周围的街坊，柔化了建筑在街道上的景致。这个平台还是车库的屋顶，同时向人行道的扩展也减少了建筑的尺寸。作为住宅的最远端，平台在人行道上方不到两米。最外围是一个花

纵向剖面图
Longitudinal Section

横向剖面图
Cross Section

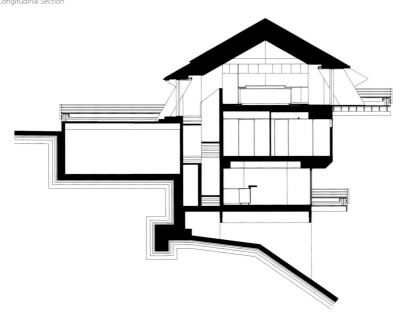

坛，小型灌木也可以使建筑看上去更小。

建筑表皮选用了户外防腐木材，经济实用、美观大方。交替的木条进一步突出了结构的形式。这种形式成功地将屋顶缘线组成了钻石的形状。垂直向上的木条强调了塔的造型，然后通过倾斜与"钻石"的边缘结合，水平的空隙间安装了玻璃。伸出东立面的壁炉外侧也是这种醒目的木条。同样出于经济考虑选择的金属屋顶也增加了立面的质感，当和木质立面结合时，金属屋顶整齐的褶皱强化了结构的几何造型。

住宅的内部和外部一样有趣。上下层通过楼梯连接，楼梯的台阶悬挂在墙壁上。台阶末端被不锈钢索串连，贯穿整个三层楼。他们营造了和外部一样的垂直感，悬臂让人联想到楼下和楼上掩映在树中的两块平台。

下层有一个娱乐室，一个浴室和悬臂平台，由滑动门进入，中间层包含3个卧室，各具特色，在建筑两端依次排开。这两层大小一模一样，经验老到的建筑师完全可以做到，但是上层提供了住宅中最动态的空间。

上层天花板的斜度和屋顶斜度一致，点缀了一系列旋转下射灯。通过这个斜天花板将外面的钻石形带入这个空间，地板上的白色扇形细节，与外立面覆层形成对比，覆层的开口与房间的主轴线垂直。壁炉是这间房间的主要亮点，独自矗立在宽阔的玻璃窗里，和房间其他地方一样也是白色，与玻璃和橡木地板形成对比。它反映了天花板的角度，但始终显示着它自身即是一个独特的元素。房间另一端一个整洁的厨房被钻石形天窗照亮。房间中间被一个滑动门隔断，向一个悬伸的平台开放，平台高耸在树冠上，提供了观看俄勒冈州的景观的神奇视角。

韦斯住宅是一个在设计方面令人鼓舞的例子，特别是受到了来自倾斜地势的制约，仍能作为一个经济住宅的事实更令人震惊。因经济原因选择的创新建筑系统实际上变成了建筑资产。外立面覆层和屋顶的结合强调了建筑的形式，内部细节使整个建筑整齐划一。罗伯特·奥赫兹用了如此别具匠心的方式处理了这个住宅设计中的困难，韦斯住宅看上去一点都不像经济型住宅。它看上去更像是一个精雕细琢的建筑设计。

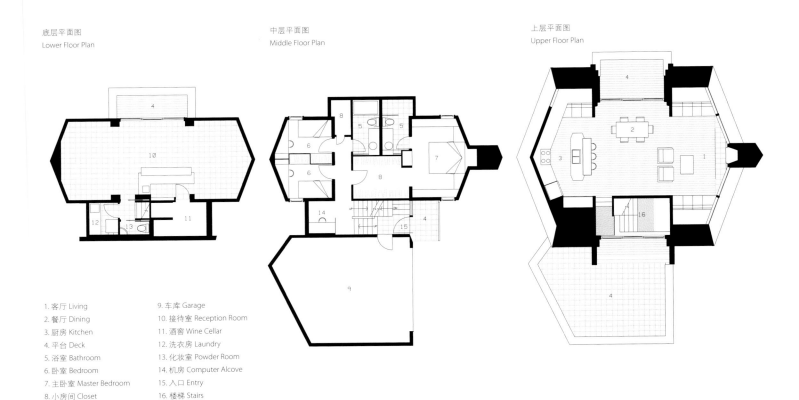

底层平面图 Lower Floor Plan
中层平面图 Middle Floor Plan
上层平面图 Upper Floor Plan

1. 客厅 Living
2. 餐厅 Dining
3. 厨房 Kitchen
4. 平台 Deck
5. 浴室 Bathroom
6. 卧室 Bedroom
7. 主卧室 Master Bedroom
8. 小房间 Closet
9. 车库 Garage
10. 接待室 Reception Room
11. 酒窖 Wine Cellar
12. 洗衣房 Laundry
13. 化妆房 Powder Room
14. 机房 Computer Alcove
15. 入口 Entry
16. 楼梯 Stairs

平台还是车库的屋顶，同时向人行道的扩展也减少了建筑的尺寸
The deck also is used to roof the garage and reduces the scale of the building by extending out toward the sidewalk

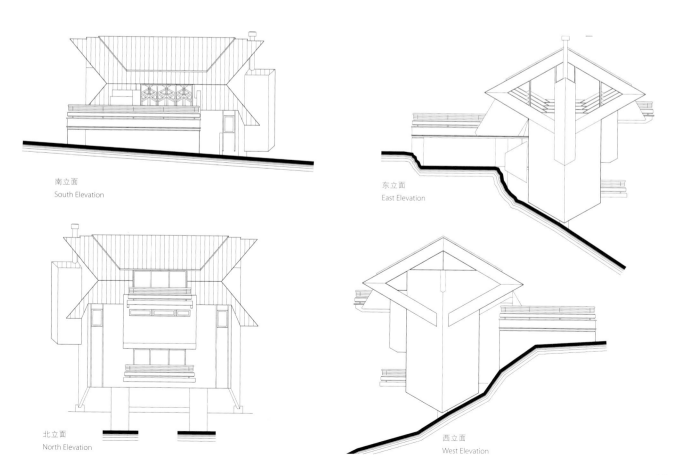

南立面
South Elevation

东立面
East Elevation

北立面
North Elevation

西立面
West Elevation

The skin of the house is constructed from outdoor treated lumber, chosen for economy, yet in its utilization provides a beautiful addition to the design. The strips of wood, used with alternating profiles are combined to form a striped elevation that helps to emphasize the forms of the structure. This is used to great success in the articulation of the diamond shape that is formed by the splitting of the eaves from the roofline. The stripes which run vertically up the building and emphasize its tower like form, are then sloped, perpendicular to the edges of the diamond, turned again, and carried through horizontally to the glazing. The fireplace, which seems to grow out of the Eastern elevation, is also accentuated by the striping effect. The metal roofing, which was also chosen for economy is also used to add to the texture of the elevation. When used in conjunction with the wood cladding system the straight folds of the metal roof help to enhance the geometric forms of the structure.

The interior of the house is equally as interesting as its exterior. The lower and upper floors are accessed via a stair case with treads cantilevered out from the wall. The stairs are suspended at their extremities by stainless steel cables that run the full height of the three story stair well. They create a vertical emphasis similar to that achieved by the exterior cladding, while the cantilever is reminiscent of the two decks that project out amongst the trees from the lower and upper floors.

The lower floor is home to an entertainment room, with a bathroom and the cantilevered deck, accessed via sliding glass doors, while the middle floor contains three bedrooms which each exhibit interesting formations in plan that are in turn expressed through the exterior. These two levels are executed with the same precision as could be expected from a skillful architect, but it is the upper floor that provides the most dynamic space in the house.

The upper floor sits beneath a raking ceiling that follows the slope of the roof, and is punctuated with an arrangement of swiveling down lights. The diamond form seen on the exterior is carried through the space by means of the raked

Located on a sloping site in hills of West Linn, a suburb of Portland, Oregon, the Weiss Residence provides an exclamation mark in the streetscape. It's strikingly rigid geometric forms and unusual detailing which accentuate its shape make for an intriguing composition. While not being an example of Robert Oshatz' most elaborate work, it may be his most admirable. Designed for Oshatz' own daughter and her husband, the Weiss Residence was built as an affordable first home for the couple and their children. Given the restrictive budget; coupled with the steeply falling lot it is astounding that such a unique piece of architecture was achieved.

The 179-square-meter building is spread over three floors giving the house a vertical emphasis and making the East and West elevations somewhat tower like, yet the building is by no means menacing from the street. The entry to the home is made onto the middle level after descending a short concrete driveway, while the upper floor is almost absorbed as the eaves slope downward to form the diamond shapes that can be seen on the side elevations. The building opens out to a large rectangular deck through sliding doors. The deck is intended to act in the same was as would a traditional porch; helping to merge the lives of the home's occupants with the rest of the neighborhood, this helps to soften the building from the street. The deck also is used to roof the garage and reduces the scale of the building by extending out toward the sidewalk. At its farthest extremity from the house the deck sits less than two meters above the sidewalk. The facade is further broken down by the planter box, adorned with small shrubs which help to reduce the scale of the building.

ceilings, and white scalloped details which grow out of the floor and collide with the exterior cladding which is brought through the openings perpendicular to the room's main axis. The fireplace provides the room's main emphasis. Standing freely amongst an expansive glass window the fireplace is finished in the same white as the rest of the room yet is isolated by glass and oak floor boards. It respectfully reflects the angle of the ceiling but ultimately remains a distinct element in its own right. At the opposite end of the room a neat kitchen is lit by a clerestory window which takes up the diamond form. The middle of the room is broken by a sliding door that opens out to a cantilevered deck which soars out amongst the tree tops giving a magical view of the Oregon landscape.

The Weiss Residence is an inspiring example of design in its own right, especially given the physical restraints of a sloping block, but the fact that it was built as an affordable home makes it even more astounding. The innovative building systems that were chosen for economy are in fact turned into architectural assets. The cladding and roofing combine to emphasis the form of the building, while the interior details bring the whole building together into one unified composition. Robert Oshatz has in this house addressed a difficult brief in such an inventive way that the Weiss Residence feels nothing like affordable housing. It feels like what it is; a skillful and well considered piece of architectural design.

Credits
Location: West Linn, Oregon, USA
Area: 179 m²

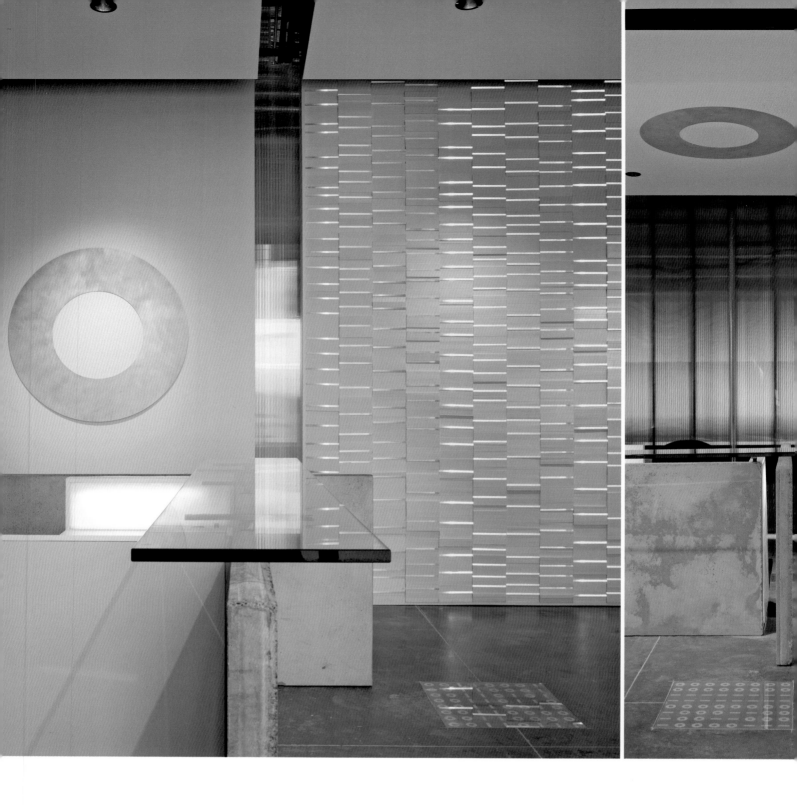

| Elliott + Associates Architects |

ImageNet Houston
独具特色的办公空间

靠近庭院的北侧立面及剖面
North elevation/section at courtyard

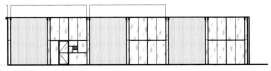

靠近庭院的东侧立面
East elevation at courtyard

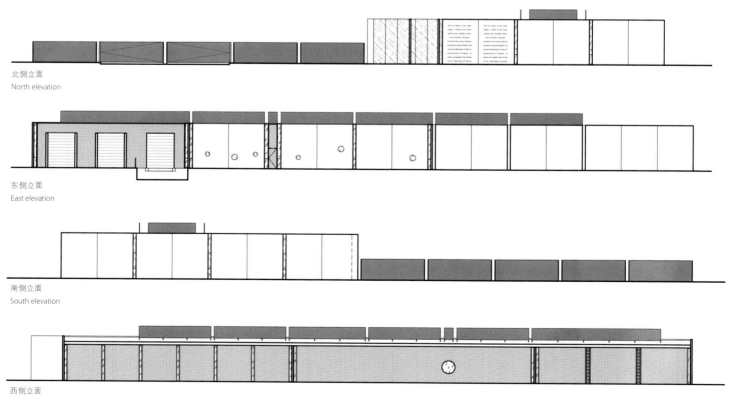

北侧立面
North elevation

东侧立面
East elevation

南侧立面
South elevation

西侧立面
West elevation

工程目标

从办公自动化到居家装潢、数码打印以及其他林林总总,美国ImageNet的所有公司拥有两个共同的特点:第一,他们为客户带来最出色、最中肯的服务,并竭诚为他们带来成功和快乐;第二,他们把为属下员工提供长期机遇、支持和顶级的办公环境作为自己的首要任务。这些员工中的绝大多数都已升任至管理层,甚至有人摇身成为拥有数个公司的老板。

设计要素

1. 这是我们为ImageNet带来的首个地面建筑。
2. 我们探索最经济、最具个性、最易扩展的楼房设计,使办公环境与公司业务有效关联。
3. 减少预算
4. 我们曾经为ImageNet设计了三个项目,分别位于三座不同的城市,每个都别具特色。俄克拉荷马市是乡间情调建筑,图尔萨城内的是装饰艺术风格,华盛顿特区的拥有勒凡特的规划,而休斯顿这个项目的特色是我们正在研究的,也想达到的效果。

项目需要

1. 设备展示厅
2. 两间视频会议室
3. 销售区
4. 管理人员办公室
5. 安全库存区
6. 修理机器的技术设备区
7. 甲板库存区
8. 大楼扩建区
9. 把环保墙纸引入设计中

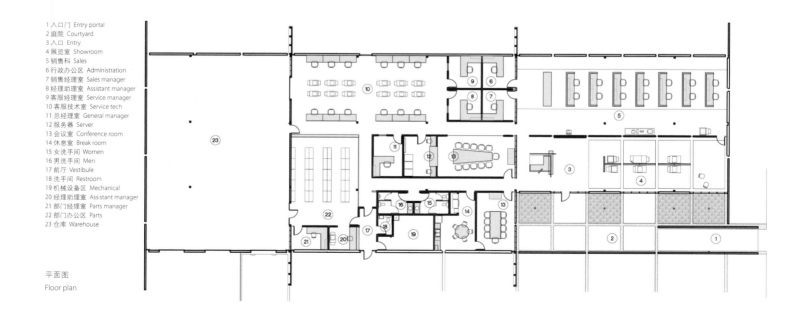

1 入口门 Entry portal
2 庭院 Courtyard
3 入口 Entry
4 展览室 Showroom
5 销售科 Sales
6 行政办公区 Administration
7 销售经理室 Sales manager
8 经理助理室 Assistant manager
9 客服经理室 Service manager
10 客服技术室 Service tech
11 总经理室 General manager
12 服务器 Server
13 会议室 Conference room
14 休息室 Break room
15 女洗手间 Women
16 男洗手间 Men
17 前厅 Vestibule
18 洗手间 Restroom
19 机械设备区 Mechanical
20 经理助理室 Assistant manager
21 部门经理室 Parts manager
22 部门办公区 Parts
23 仓库 Warehouse

平面图
Floor plan

Project Goals

From office automation to home construction to digital printing and beyond, the companies of ImageNet America share two common characteristics. First, they exist to serve customers with excellence and devotion and contribute to their success and happiness. Second, the key to accomplishing that is providing long-term opportunity, support, and a great working environment for their associates, many of whom have risen to management and even ownership roles of the various companies.

Key Points as Presented to the Client

1. This is our first ground-up project for you.

2. We are looking for economy, relevance to your business, a memorable image/personality and expandability.

3. Think of this as a low budget project.

4. We have designed 3 projects for you to date with each in a different city. Each location has its unique characteristics. Oklahoma City is home, Tulsa has Art Deco, Washington D. C. has L'Enfant's plan and Houston has characteristics that we have studied and want to respond to.

Program Requirements

1. Equipment showroom

2. 2 video conference rooms

3. Sales area

4. Management offices

5. Secure inventory area

6. Technical support area for machine repair

7. Inventory warehouse with dock.

8. Building expandability

9. Incorporate the signature recycled "paperwall" into the project

Credits

Location: Houston, USA
Area: 1,254 m²
Completion Date: 2010
Project Team: Rand Elliott, FAIA
Michael Shuck, Assoc. AIA, Project Manager
Brian Fitzsimmons, AIA
General Contractor: Mission Constructors, Inc.
Civil Engineer: Cobb, Fendley & Associates, Inc.
Structural Engineer: Haynes Whaley Associates, Inc.
MEP: E/B/E, Inc.
Landscape Architect: Wong & Associates, Inc.
Client: ImageNet Office Systems
Photographer: Scott McDonald, Hedrich Blessing
Awards: Chicago Athenaeum, International Architecture Award; Texas Society of Architects Design Award;
AIA, Central States Region, Merit Award;
Association of General Contractors, Houston Chapter, Apex Award

拱廊的飞檐造型是设计的一大亮点
Design of the curved fins on the shopping arcade is a highlight

HIMEJI FORUS | 飞檐魅影 | ITO MASARU DESIGN PROJECT / SEI |

这几年该店设计师参与了AEON（永旺）集团的"姬路FORUS"的重建工程，内容是对外观在内的整个店铺环境进行改造。十多年前设计师为"ISSEY MIYAKE MEN"服装品牌设计了姬路FORUS店，那之后和FORUS的一系列合作都是以"姬路店"为契机。

接到这次的设计邀请后设计师到现场进行了视察。时隔这么久他发现以FORUS这个城市时尚象征为中心的曾经热闹繁华的Omizo街头，行人已变得稀疏零散。

因此，为了让人们远远地从主干道MIYUKI大街就能明显注意到FORUS并被其吸引，设计必须要考虑加强视觉效果。但是，由于这个建筑还留有旧的拱廊，即使使其外观从远处就能给人以视觉冲击，但一旦走近，拱廊就会遮住视线，不能充分吸引人们。鉴于此，设计师选择通过改造拱廊来加强视觉感受。

经过多次设计讨论会后，最终决定在近100米长（包括东馆）的拱廊上做飞檐造型，使其具有延续性，并在缝隙中安装照明灯，通过照明灯投下的阴影赋予其独特的造型和富有创意的形象，这样从质感与现代感上都做到了与城下町的氛围相吻合。

另外，由于拱廊外侧是出租商铺，所以设计必须能够具有吸引人们到内部的魅力。因此，设计师在东馆尽头的楼梯井处安装了大型吊顶灯，创造了一个即使远观也足以让人停留欣赏的空间。不仅如此，墙壁上还安装了由单个FORUS标志拼成的图案发光体，让亮度增加，这样从外面的街道看起来也非常显眼。

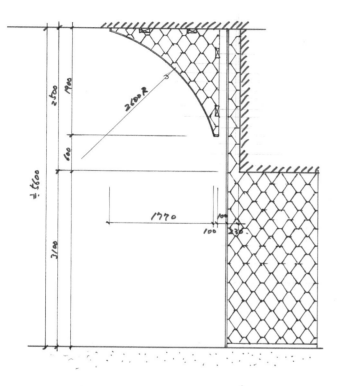

大型吊灯为空间增添了许多设计感
The pendant light is a highlight in the space

In recent years, the designer has been working on designing the whole environment of Himeji FORUS (AEON group) including exterior of the building. This time he participated in the renovation project. Since the designer designed for ISSEY MIYAKE MEN in Himeji FORUS as a tenant's shop designer more than 10 years ago, he has been designing for other FORUS buildings as well. After receiving the offer, the designer visited the site for the first time in years.

There used to be many people in Omizo Street centered on FORUS as a symbol of urban fashion in Himeji, but there is much less people in the street now. Therefore, the building needed to have a clear impact on the people in the distance who walk in Miyuki Street, the main street of Himeji. However, even if the exterior of the building can give strong impression on the people in the distance, the old shopping arcade of the building will be an obstacle in the close distance. So the designer decided to create the impressive arcade, apart from the building.

After a few design meetings, he decided to install curved fins on the about 100m long shopping arcade including the East Building to produce continuity, and set indirect lightings between the fins to generate shading to highlight unique form and give creative image. This work produced texture and modernness appropriate for Himeji, the castle town.

The eye-catching inside part of the arcade was exclusively occupied by some specific tenants. Therefore, something special to lead visitors further into the arcade was necessary. The big pendant lights were installed on the open ceiling zone at the end of the East Building to create the space where visitors will be interested in even from the distance. FORUS's icon shaped lights were also installed on the wall to produce more brightness so that people on the exterior sidewalk can be impressed when they peek inside.

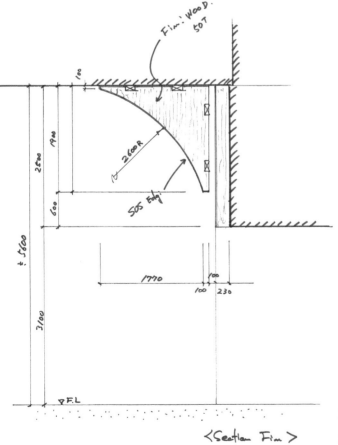

<Section Fin>

Credits

Location: Hyogo, Japan
Area: First floor of west- 220.5m², Second floor of west- 212.77m², Third floor of west-189.06m²
Construction Date: September to November, 2011
Designer: Masaru Ito
Photographer: Nacasa & Partners Inc.
Opening Time: 10 am-8 pm
Main parts: Shop 87 Service area 2 Restaurant 2 Other 1

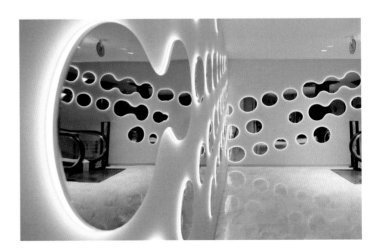

空间纵深感极强，显示出品牌的高贵与典雅
The long and deep space shows elegance of the brand

| ITO MASARU DESIGN PROJECT / SEI |

Roar (Daikanyama)
暗黑潮流——Roar品牌旗舰店

从涩谷驾车前往代官山需沿明治街前行，在Namiki-bashi右转，来到八幡街，之后穿过位于新宿的大型陆桥。
桥梁终点处便是所谓的"代官山"，Roar品牌旗舰店位于代官山区右侧边缘处。
用"空中铁匠"乐队的歌曲名"Living on the Edge"来描绘该店再贴切不过了。这是Roar品牌的首家旗舰店。Roar以其交叉枪logo而闻名，近年来尤其在东京广受好评，在海外也声誉颇佳。
店面的主题外形是一件破损的Roar牌T恤衫，上面的交叉枪logo用几千颗施华洛世奇水晶点缀而成。

设计师基于该理念对于设计过程进行数次讨论，最终在共同兴趣的影响下，他们决定将其建成以停车库为主题的艺术空间。
沿着"Roar"店面标记寻去，穿过黑色金属大门，再穿过"小径"后，来到移动玻璃门前，"小径"由黑色印花钢板装饰，而移动玻璃门由黑色玻璃铸造而成。
在穿过移动玻璃门之后，是一条黑暗、弧形的狭窄地道，通过上方灯具照明。穿过地道后，终于可以感受Roar品牌店的氛围了。悬于顶部的大量吊灯仿佛在欢迎进店的顾客。在机械化时代，这些灯具隐藏在幕后对道路进行照明，如今已不再使用。它拥有工业化设计的美丽样式，从某种意义上

讲它是店面的中心标志。
这条狭长的空间约有100平方米，呈半黑暗式。
由砖瓦砌成的墙壁看上去仿佛历经沧桑，而16.5厘米钢网的百叶窗样式平坦地悬挂在各处，互相对称。
百叶窗的空隙处便是悬挂区，而经久耐用的脚手板则附于空间后方。同样的脚手板还用于地板安装，显示了材料的质感。所有脚手板都用清晰的玻璃覆盖，从而形成间接光照。显然所有耐用的板会使大气变得干燥多尘。然而，对于玻璃下的脚手板而言，如同美术馆一般，散发出整洁和高贵的气质。
这样的布置绝对复古又现代。3米高的桌子乃手工铸成，其材料曾用于设计法老陵墓的中心处。
空间尾端的铁门上配有数千把复古钥匙，如同"潘多拉的魔盒"一般，我们无法进入这如同寺庙一般的塔楼。
这些经久耐用的构件，专业的技术，以及精美的技艺构成了"Roar"的品牌文化，值得细细品味。
品牌精神融入了产品之中，并浸透了设计师自己的生活体验，营造出独一无二的"Roar"品牌氛围。

洗手间墙壁：皮质所造，瓷砖表面
Toilet wall: leather molded surface tile

吊灯：公用通道灯具
Suspention light: highway light

地板：砂浆表面
Floor: motar trowel finish

桌子：钢架 + 老式脚手架板 + 玻璃
Table: mill saled steel frame + vintage scaffolding board + clear glass

窗帘：皮质
Curtain: leather curtain

平面图
Plan

火山灰：钢架 + 复古表面
Trass: steel frame + aging finish

墙：砖瓦 + 玻璃屏
Wall: brick tile + glass screen

地板：老式脚手架板 + 硬质玻璃
Floor: vintage scaffolding board + toughened glass

皮沙发：Roar 品牌原生态
Leather sofa: "Roar original"

货架：钢质清晰表面
Shelf: steel clear finish

柜台：黑色不锈钢，镜式表面
Counter: black stainless steel mirror finish

移门：老式镀锡薄钢板
Sliding door: vintage tin panel

幽暗的空间透露出Roar品牌暗涌的时尚潮流
The space of quiet and dark shows the fashionable style of Roar

Drive down for Daikanyama from Shibuya, and on Meiji-street, turn right at Namiki-bashi onto Hachiman Street, and then go over the large land bridge of JR.

The area from the end of the bridge is so-called Daikanyama. The flagship shop of Roar is located right on the edge of Daikanyama

The situation is exactly like the one described in Aerosmith's song "Living on the Edge".

This is the first flagship shop project of "Roar". "Roar" is famous for the crossed guns motif and is highly acclaimed especially in Tokyo and also in overseas in recent years, and now mounts an exhibition in Paris.

The request from Mr. Hamanaka, a friend and designer, is to construct the shop thoroughly with the concept as the Roar's damaged T-shirt with the crossed guns motif printed with thousands of Swarovski (This request largely comes form Mr. Hamanaka's stance of creating clothes).

Mr. Hamanaka and the designer conducted the design session a few times with the concept, and finally, with the influence of our mutual interest, we decided our theme—to create the extremely artistic space with the motif of garage.

Follow the small shop-sign of "Roar" and go through the black expand-metal gate, and reach the slide-door made by black glass after going through the approach decorated with nothing but series of the steel panel finished with black mill scale on the left side. After going through the auto slide-door, there will be a dark, curved, narrow tunnel illuminated only with upper-light. After the tunnel, we finally can feel the space of roar. The huge mass of highway light dangled from the ceiling welcomes the guest first. It used to light up the road for decades as a behind the scenes of the Machine Age and now is retired. It has a beautiful form with industrial design, and exists as a central icon of this shop in a way.

The long and narrow square shaped space of about 100 m² or less is in semi-darkness in whole.

The walls consist of brick tiles finished with aging treatment, and the shutters numbered with 5-span stencil are placed evenly and in symmetry on each side. The each shutter span is the hanging space, and the scaffold board with the history in well-used condition is attached on the back of the space.

The similar scaffold boards are used for the floor, and materialness is expressed. All of them are covered with the clear glass installed with the indirect lighting. Normally those well-used boards emit rough and dusty atmosphere. However, the atmosphere seems to be the gallery of the boards beyond the glass, and the boards are emitting clean and luxury aura.

It is truly out of respect for the historical things, and also the expression of praise from us with a feeling of awe to such things.

The 3 m table handmade by the authority of ironwork occupies the center of the space as the coffin of Pharaoh.

At the end of the space, the steel gate with thousands of antique key like a Pandora's box we never should open towers like a temple.

The taste of these well-used things, the expert's skill, and the cutting edge of technique are all part of spirit of the brand "Roar".

They blend in and melt down with the products infused with the life by same designer, and create an unprecedented original atmosphere of "Roar".

灯光映在墙壁上，依稀看见空间工业化时代的美丽
We can see the beauty of industrial times from the light on the wall

| Stefano Tordiglione Design |

apple & pie Children-shoe Boutique
apple & pie 童鞋专卖店

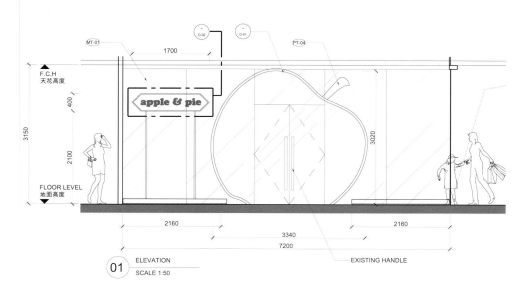

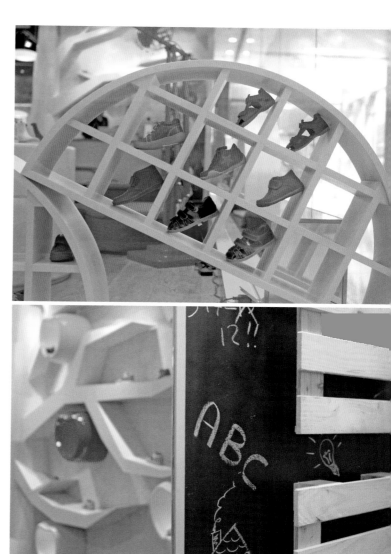

童趣、典雅、实用是意大利室内设计事务所Stefano Tordiglione Design为童鞋品牌apple & pie于香港One Island South新专卖店塑造的三个截然不同的感觉。

走进apple & pie童鞋专卖店就像走进了另外一个世界——门口的巨型红苹果隐约地暗示了店内的许多趣味。受到品牌名字以及其半苹果半馅饼商标的启发，Stefano Tordiglione Design的空间概念融合了苹果代表的健康元素以及馅饼所体现的趣味性。苹果的健康意味通过环保物料和有益儿童的物料体现出来。店中运用了大量的木质材料，并尽量减少塑料材料的使用。馅饼的趣味就在空间异想天开的设计中展示出来，包括了苹果形状大红沙发和充满想象力的墙身展示。另外，每个设计元素都充分发挥实用功效。红苹果沙发内含收纳空间。生动的苹果树上长满红色和白色的果实，可爱的馅饼造型货架放在孩子伸手可及的地方，这些元素都充分发挥展示

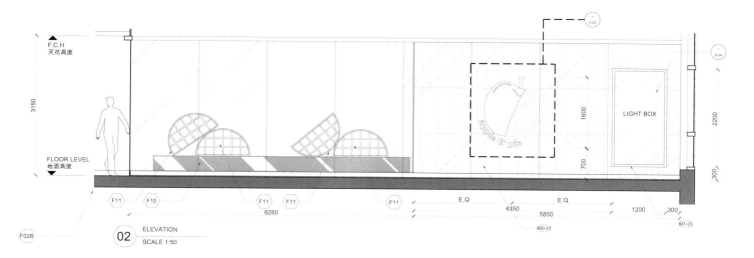

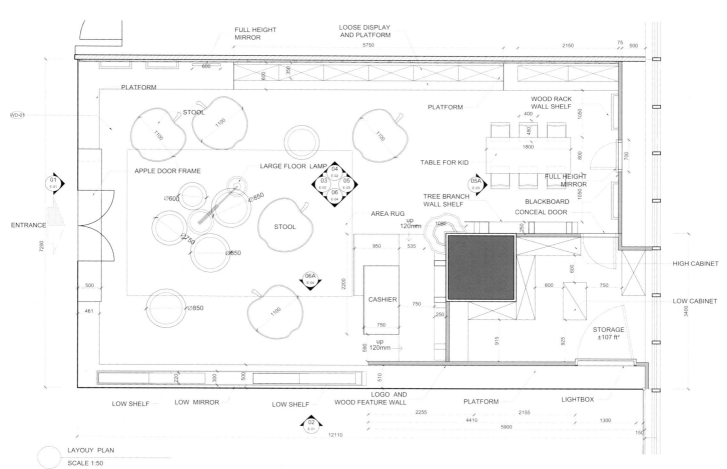

LAYOUY PLAN
SCALE 1:50

产品的功能。橱窗里，馅饼造型货架让品牌的名字与商标结合，游人在店外已能观赏到欧洲各个品牌的鞋子。

不论客户的年龄如何，Stefano Tordiglione Design都充分考虑到空间的布局。对于年轻的小客人，店的后面设计了一面巨形的黑板，让孩子觉得熟悉舒服。黑板旁边放了一张小桌子，小孩子可以在试鞋的同时休息和玩游戏。但是，店的设计并不只针对于孩子。小桌子旁边的Kartell椅子、天花板挂下来的Ethel吊灯，以及Anglepoise设计的巨形红台灯，通通都体现了店铺典雅高贵的气质。整个店由柔顺的曲线和简洁的直线组合而成，板也用上了柔和色调的木板，再加上相似和对比颜色的交替使用，例如薄荷绿和米白色的柔和，相对于艳红和翠绿的对比，都让店铺轻松地游走于小孩和大人的世界。

apple & pie童鞋专卖店创造了设计的先河：这是一个孩子可以自由挑选自己鞋子的世界，整体的环境开心欢乐，玩味十足。通过默默地耕耘，Stefano Tordiglione Design创造了美好的环境，让孩子的购鞋需要得到充分满足。

小桌子旁边的Kartell椅子、天花挂下来的Ethel吊灯，以及Anglepoise设计的巨形红台灯，通通都体现了店铺典雅高贵的气质
The iconic Kartell chairs surrounding the low table, the Ethel lighting hanging from the ceiling above, and the Giant Red Lamp designed by Anglepoise are design features which lend a sophisticated and elegant air to the store

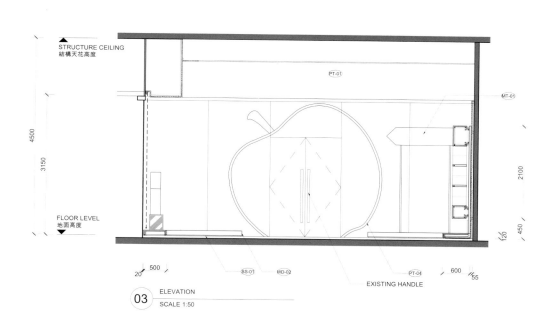

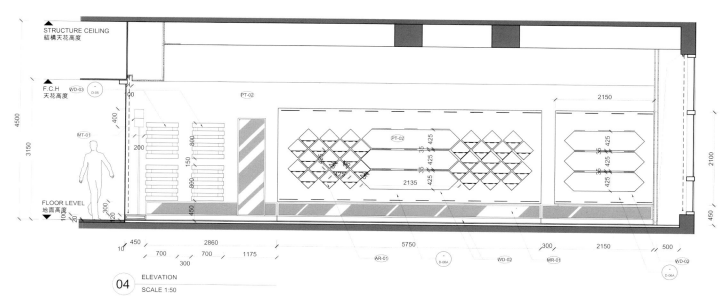

店中运用了大量的木质材料，并尽量减少塑料材料的使用
Child-friendly materials are focused on wood as opposed to plastic for the furnishings.

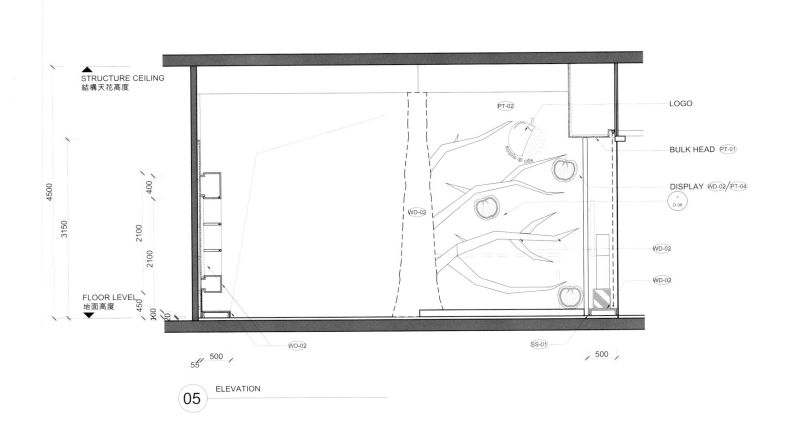

05 ELEVATION

Playfulness, elegance and practicality come together at One Island South's new apple & pie Boutique, the latest creation from Hong Kong-based Italian interior design and architecture firm Stefano Tordiglione Design.

Walking into children's shoe store apple & pie is like entering another world— the giant apple that crowns the doorway hinting at the many delights that lie within. Inspired by the ethos behind the brand's name and its half apple-half pie logo, Stefano Tordiglione Design's concept combines the wellbeing elements represented by the apple with the more playful pie. The former is reflected in the use of environmentally- and child-friendly materials with a focus on wood as opposed to plastic for the furnishings, while the latter can be seen in the whimsical interior design which ranges from bright red apple-shaped sofas to imaginative wall displays. There is also practicality behind each well-thought out element. The seating hides storage space, while a lively tree design on one wall with white and red apples hanging from its branches, and elsewhere pie-shaped lattices and mounted fruit palettes, offer ideal shelving opportunities. In the windows, semi-circular pie-like features bring the logo and brand name full circle while providing window-shoppers with a taste of the various European shoe brands that can be found within.

Throughout the space Stefano Tordiglione Design has considered the experience of its clientele, young and old. For children coming to try and buy shoes, the back of the store offers a table at which they can sit and play between fittings, not far from a wall that provides familiarity through its giant blackboard design. Yet the focus is not solely on a positive experience for children. The iconic Kartell chairs surrounding the low table, the Ethel lighting hanging from the ceiling above, and the Giant Red Lamp designed by Anglepoise are design features which lend a sophisticated and elegant air to the store. Coupled with a combination of smooth curves and clean lines, above a warm wood-lined floor and with a color palette that blends bold reds and vivid greens with a calming mint and light beige, the store effectively and effortlessly moves between the distinct worlds of parents and children.

The flagship apple & pie store sets a precedent—it is a place where children can enjoy choosing shoes that are displayed in a fun and enticing way in cheerful, relaxed surroundings. Through the fruits of its labor, Stefano Tordiglione Design has created the ideal environment in which children's shoe shopping needs can be met.

Credits

Location: One Island South, Hong Kong, China
Completion Date: August 2012
Area: 85 m²
Lighting: Ethel, Anglepoise
Furniture: Kartell
Construction: Anzac
Text: Rachel Duffell
Shop Design Studio: Stefano Tordiglione Design Ltd
Logo Design Studio: Stefano Tordiglione Design Ltd
Chief Designer: Stefano Tordiglione
Photos: Stefano Tordiglione Design Ltd

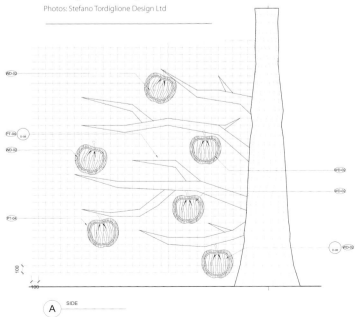

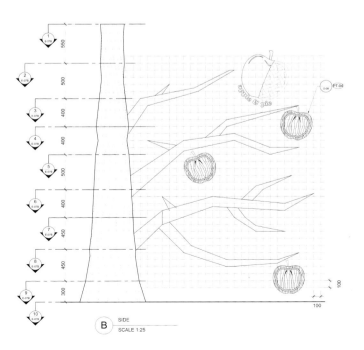

Burdifilek室内设计事务所设计的布朗托马斯珠宝大厅完美地捕捉了产品的精髓,引领顾客进入极致奢华的殿堂。布朗托马斯专注于生产名贵珠宝、名表和高档工艺品,这个标志性的空间参照了布朗托马斯品牌经典的现代主义设计,强化了这个百货商店作为爱尔兰奢侈品购物天堂在都柏林心脏地带的地位。

为了提升奢侈品的购物体验,设计中对细节的关注徐徐透出一种微妙的优雅。作为中央展厅,优雅的雕饰鼓励顾客驻足观光,不经意间引领顾客穿梭于空间中。香槟色的斯塔菲喷砂玻璃底座上,半圆形的展柜悬浮于带有褐色、奶油色和咖啡色的奶油状斑驳的蜜色大理石地板。定制的牡蛎色绒布作为珠宝低调的背景,很好地将珠宝推向前台。

周围是高反射性的双色玻璃,使空间看上去熠熠生辉,透明的玻璃制造出悬浮的无形展柜,可以从像卡地亚和蒂梵尼这样的邻近品牌店望见这里的橱窗。

珠宝的光泽在周围凹凸的空间里被进一步反射。多伦多艺术家丹尼斯·林设计的抛光镍雕塑使室内充满了活力。从地板伸向天花板,金属棒优雅地凝固在空间里。有色古镜表面进一步反射了抛光镍雕塑,最大化了它的视觉效果。

Burdifilek 股东保罗·弗莱克评价道:"这个奢华大厅在都柏林旗舰店里占据独特的地位,反映了布朗托马斯对最高级别零售场所的一贯追求。"它是继2006年获奖的女式设计师系列和鞋品商店,以及2008年的男人世界之后又一Burdifilek与标志性的零售商合作的室内设计成功范例,每一处都是国际认可的最佳购物圣地。

Burdifilek设计的布朗托马斯奢华大厅已经被列在不少于四个独一无二的竞赛名单上:加拿大最佳室内设计奖,安大略注册室内设计师协会奖,英国国际地产大奖和英国《零售周刊》的零售室内设计奖。

布朗托马斯珠宝店巧妙地运用了朴素的颜色和反射物质
The Brown Thomas Jewellery Department is a study in understated coloration and reflective materiality

Brown Thomas Jewellery
布朗托马斯珠宝店

| Burdifilek Interior Design

抛光镍雕塑是室内的焦点
Hand-articulated polished nickel sculptures provide a distinct point of memory within the environment

Beautifully capturing the essence of the product it houses the Burdifilek-designed Brown Thomas Jewellery Hall transports shoppers into a world of refined luxury. Dedicated to fine jewellery, watches and luxury gifts, the signature space references the sophisticated modernism of the Brown Thomas brand; strengthening the department store's positioning as Ireland's luxury shopping destination in the heart of Dublin.

Focusing on elevating the luxury shopping experience, the attention to detail within the design instills within it a sense of nuanced elegance. Concieved as central piazza, gracefully curved displays encourage meandering exploration and effortlessly lead shoppers through the space. Cantilevered over sandblasted champagne colored Starfire glass bases, semi-circular fixtures appear to float above honey marble flooring in mottled creamy shapes of taupe, cream and cafe-au-lait. Custom oyster-colored suede displays provide an inconspicuous backdrop for the jewellery itself, allowing the product to take center stage.

The perimeter is sheathed in a veil of subtly reflective dichroic glass; revealing a shimmering gold-hued iridescence. Invisible floating vitrines are suspended on the modulated transparent glass, allowing for a glimpse through adjacent shop-in-shops for brands like Cartier and Tiffany & Co.

The materiality of the product is further reflected in recurring sculptural

调制的透明双色玻璃墙可以看见旁边的品牌店
Modulated transparent dichroic glass walls allow for a glimpse through to adjacent shop-in-shops

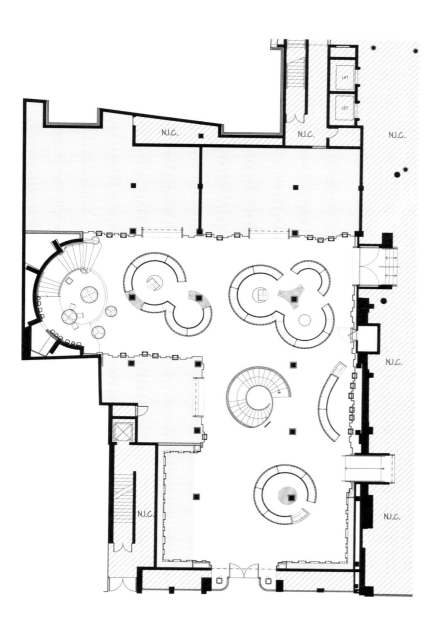

installations. Hand-articulated in polished nickel, Bertoia-inspired installations by Toronto-based artisan Dennis Lin infuse the interior volume with a kinetic energy. Extending from floor to ceiling, rods appear to be elegantly captured in suspended animation. Columns clad in tinted antiqued mirror further reflect the polished nickel sculptures, maximizing its visual impact.

"The Luxury Hall has a distinctive voice within the Dublin flagship, and reflects Brown Thomas' continued commitment to the highest level of retailing," comments Paul Filek, Burdifilek's Managing Partner. The interior design concept carries on the success of Burdifilek's ongoing collaboration with the iconic retailer, having also designed the award-winning Women's Designer Collections and Shoe Departments in 2006, and the Men's World in 2008, each contributing to the retailer's international recognition as a premier shopping destination.

Burdifilek's design for the Brown Thomas Luxury Hall has been shortlisted in no less than four unique competitions: Canadian Interiors Best of Canada Awards, Association of Registered Interior Designers of Ontario ARIDO Awards, UK's International Property Awards as well as Retail Week's Retail Interior Awards in the UK.

Credits

Location: Dublin, Ireland
Completion Date: 2009
Floor Area: 539m²
Client: Brown Thomas
Photographer: A Frame Inc. – Ben Rahn

NEWS 新闻

全球顶级建筑事务所墨菲/扬（Murphy/Jahn)正式更名为JAHN

2012年10月26日，事务所首席执行官、世界建筑大师赫尔穆特·扬（Helmut Jahn）以亲笔手稿宣布全球顶级建筑事务所——墨菲/扬（Murphy/Jahn)正式更名为JAHN。与此同时，建筑师弗朗西斯科·冈萨雷斯-普利多（Francisco Gonzalez -Pulido）晋升为事务所总裁，与赫尔穆特·扬共同领导事务所。他们将一起带领JAHN建筑事务所，在世界建筑界建立新的卓越标杆。

"正如我们事务所的新名字，JAHN将超越赫尔穆特过去的成就，并为未来的卓越建筑设定最高标准。"11月16日，冈萨雷斯-普利多在视察中国项目时接受媒体采访并表示，"JAHN这个名字不但诠释着赫尔穆特对这支优秀团队所倾注的价值观，还激励着我们继续挑战行业的极限。目前，我们在亚洲地区非常活跃，业务包括中国、韩国和日本等地的大型在建项目。同时，为了适应中国这一地区近年来快速发展的项目需求，我们于2011年在上海设立了东亚地区的首个办事处。我们希望这除了能更好地服务中国境内的多个项目外，还能将JAHN的创新建筑理念引入中国，以期为中国带来更多具有前瞻意义的建筑作品。"

时代楼盘第七届金盘奖隆重揭晓 万科再次独占鳌头

2012年11月9日，由《时代楼盘》主办，唐艺设计资讯集团承办的第七届金盘奖在上海紫金山大酒店隆重拉开帷幕，时代楼盘金盘奖活动是楼盘设计类的评选活动，以"权威、专业、公平、公正、公开"为宗旨，以"人文、科技、艺术、创新"为标准，从科技和艺术的高度提升我国住宅的品质。

本次活动得到了全国60余家专业媒体的追踪报道，搜狐焦点网进行全程图文和视频直播，《时代楼盘》和活动官方网站金盘网、官方微博新浪@时代楼盘@金盘地产资讯进行全程报道，引发业界广泛关注和热议。

活动由时代楼盘执行主编杜全利，上海美地行营销策划有限公司合伙人、策划总监周海云主持。15位专家评委团通过投票方式，评选出年度年度最佳商业楼盘、最佳写字楼、年度最佳公寓、年度最佳别墅、年度最佳样板房、年度最佳酒店、年度最佳保障房、年度最佳综合楼盘等八个类别的优胜项目。

此次活动由唐艺设计资讯集团总经理康建国先生致开幕词，他表示，时代楼盘第七届金盘奖评选是在房地产市场调控大背景下举办的。本届"金盘奖"评选主题本着"尊重城市、尊重设计"的原则，尊重社会价值的肯定，对于不同人需求的尊重，对于不同城市历史发展的尊重，尊重不同人文的特性，尊重每个居住人的感受和尊重自然。

在历时将近九个小时的评选过程中，各位评委嘉宾进行了激烈的讨论，无论是对参选项目还是对于分类方式，都开诚布公地提出了见解，并对写字楼、商业楼盘、住宅等未来的趋势提出了个人的预测，并希望时代楼盘能够提供更广阔的平台，为行业的发展能够做出更多的努力。